REA: THE TEST PREP AP TEACHERS RECOMMEND

2nd Edition

AP* BIOLOGY
CRASH COURSE®

Michael D'Alessio, M.S.

Lauren Gross, Ph.D.

Jennifer Guercio, M.S.

Research & Education Association

Visit our website at: www.rea.com

Research & Education Association
61 Ethel Road West
Piscataway, New Jersey 08854
E-mail: info@rea.com

AP BIOLOGY CRASH COURSE®

Published 2014

Copyright © 2013 by Research & Education Association, Inc.
Prior edition copyright © 2011, 2010 by Research & Education
Association, Inc. All rights reserved. No part of this book may
be reproduced in any form without permission of the publisher.

Printed in the United States of America

Library of Congress Control Number 2012953516

ISBN-13: 978-0-7386-1099-3
ISBN-10: 0-7386-1099-2

Cover image: ©iStockphoto.com/LeoGrand

D14

AP BIOLOGY CRASH COURSE
TABLE of CONTENTS

PART I INTRODUCTION

PART II EVOLUTION

PART III CELLULAR PROCESSES: ENERGY AND COMMUNICATION

ABOUT THIS BOOK

REA's *AP Biology Crash Course* is the first book of its kind for the last-minute studier or any AP student who wants a quick refresher on the course. The *Crash Course* is based on the latest changes to the AP Biology course curriculum and exam.

Our easy-to-read format gives students a crash course in Biology. The targeted review chapters prepare students for the exam by focusing on the important topics tested on the AP Biology exam.

Unlike other test preps, REA's *AP Biology Crash Course* gives you a review specifically focused on what you really need to study in order to ace the exam. The review chapters offer you a concise way to learn all the important facts, terms, and biological processes before the exam.

The introduction discusses the keys for success and shows you strategies to help you build your overall point score. Parts 2 through 5 are made up of our review chapters. Each chapter presents the essential information you need to know about biology.

Part 6 focuses on writing the essays for the AP Biology exam and the science practices that will be tested on the exam. Part Six concludes with explanations of the 13 AP Biology Labs.

No matter how or when you prepare for the AP Biology exam, REA's *Crash Course* will show you how to study efficiently and strategically, so you can boost your score!

To check your test readiness for the AP Biology exam, either before or after studying this *Crash Course*, take REA's **FREE online practice exam**. To access your practice exam, visit the online REA Study Center at *www.rea.com/studycenter* and follow the on-screen instructions. This true-to-format test features automatic scoring, detailed explanations of all answers, and diagnostic score reporting that will help you identify your strengths and weaknesses so you'll be ready on exam day!

Good luck on your AP Biology exam!

ABOUT OUR AUTHORS

Michael D'Alessio earned his B.S. in Biology from Seton Hall University, South Orange, New Jersey, and his M.S. in Biomedical Sciences from the University of Medicine and Dentistry of New Jersey. He has had an extensive career teaching all levels of mathematics and science, including AP Biology. Currently, Mr. D'Alessio serves as the Supervisor of the Mathematics and Business Department at Watchung Hills Regional High School in Warren, New Jersey.

Lauren Gross earned her B.S. in Biology from Dickinson College and her Ph.D. in Plant Physiology from Pennsylvania State University. She currently teaches AP Biology to homeschooled children in the United States and abroad for Pennsylvania Homeschoolers, where she is also a home education evaluator. As an assistant professor at Loyola College in Maryland, Ms. Gross taught various biology, genetics, and botany courses.

Jennifer C. Guercio earned an M.S. in Molecular Biology with a concentration in neuroscience from Montclair State University, Montclair, New Jersey. For the past several years, she has been doing research in neuroscience as well as teaching academic writing at Montclair State University. Ms. Guercio attended North Carolina State University as a Park Scholar where she earned her B.A. and M.A. degrees.

ACKNOWLEDGMENTS

In addition to our editor, we would like to thank Larry B. Kling, Vice President, Editorial, for his overall guidance, which brought this publication to completion; Pam Weston, Publisher, for setting the quality standards for production and managing the publication to completion; Diane Goldschmidt, Senior Editor, for editorial project management; Alice Leonard, Senior Editor, for preflight editorial review; and Weymouth Design and Christine Saul, for designing our cover.

We would also like to extend special thanks to Ernestine Struzziero of Lynnfield High School, Lynnfield, Massachusetts, for technically reviewing the manuscript, Marianne L'Abbate for proofreading, and Kathy Caratozzolo of Caragraphics for typesetting this edition.

FOREWORD

The AP Biology examination will be a measure of how much you have learned throughout the year in your AP Biology class. This *Crash Course* has been written specifically to help you achieve success on the exam. It covers all the material and themes—the "Big Ideas"—that are stressed throughout the revised AP Biology curriculum. There is also additional material on the inquiry-based Laboratory Investigations, and pointers on how to write a comprehensive essay for the free-response section.

REA's *AP Biology Crash Course* will give you an idea of how well prepared you are before taking the exam. You will be able to determine for yourself which concepts will require additional study. Keep in mind this is *not* a textbook, but rather a unique way to approach your preparation for the exam. The material is presented in a convenient outline format and includes numerous illustrations to help you better understand the material.

The AP Biology examination is a cumulative test based upon a year-long course of study. With this *Crash Course*, you're well on your way to achieving success on the AP Biology exam.

Best effort,

E. A. Struzziero
AP Biology Teacher
Lynnfield High School
Lynnfield, Massachusetts

PART I

INTRODUCTION

Keys for Success on the AP Biology Exam

I. Using the *AP Biology Crash Course* to Prepare for Success

Beginning with the May 2013 test administration, the AP Biology exam underwent a radical change. Instead of focusing on broad topics (cells, evolution, etc.), the revised AP Biology exam tests students on their critical-thinking abilities and performance in inquiry-based labs. Don't worry—these changes are covered in this book.

This *Crash Course* is based on a careful analysis of the revised AP Biology course curriculum and exam format. Parts 2–5 provide you with a detailed review of each of the topics from the AP Biology syllabus, in the same order as the syllabus itself. Part 6 covers everything you need to know about the AP Biology labs, science practices, and writing an essay.

This *Crash Course* contains all the information you need to know to earn a score of 4 or 5. Use it as a supplement to your coursework and as a final review in the last few weeks before the exam.

1. The Content of the Advanced Placement Biology Examination

The revised Advanced Placement Biology course is focused on building students' understanding of biological concepts and developing their reasoning skills in a scientific laboratory setting. The AP Biology curriculum is based on 4 Big Ideas that are designed to help students understand core scientific principles and other biological concepts. The course also includes 13 inquiry-based laboratories. The labs test AP students' inquiry skills and encourage them to think like scientists.

The AP Biology course is the equivalent to a two-semester college-level introductory biology course. In order to succeed on the exam, students need to master the key concepts that make up the 4 Big Ideas and apply these concepts to various situations in a traditional test format. The 4 Big Ideas are:

Big Idea 1: Evolution—The evolutionary process is responsible for the diversity of life.

Big Idea 2: Cellular processes: energy and communication—Biological systems use molecular building blocks and energy to maintain homeostasis, reproduce, and grow.

Big Idea 3: Genetics and information transfer—Living systems retrieve, transmit, store, and respond to information essential to life processes.

Big Idea 4: Interactions—Biological systems interact and possess complex properties.

2. The Structure of the Exam

The AP Biology exam is made up of two sections: multiple-choice and free-response. Each section includes questions that test students' understanding of the 4 Big Ideas.

The exam is 3 hours in length. It is comprised of 63 multiple-choice questions, 6 grid-in questions, 6 short free-response questions, and 2 long free-response questions. The grid-in questions require that you perform calculations with a calculator (graphing calculators are not allowed) and fill-in the bubble with the value. You will not be given any answer selections for the grid-in questions.

Section I: Multiple-choice = 50% of the exam grade

Parts A and B: 90 minutes, 63 multiple-choice questions; 6 grid-in questions

The multiple-choice questions will consist of four answer options (A through D) instead of the five answer choices that have historically characterized the exam. As on previous AP Biology exams, your score will be based upon the number of correct responses you give. No scoring penalties are imposed for incorrect or unanswered questions.

The 6 grid-in questions will test the students' science and mathematical skills. Students will be required to calculate the correct answer for each question and fill it in on a grid on the answer sheet.

A sample appears below:

Integer answer 502	Integer answer 502	Decimal answer −4.13	Fraction answer −2/10

Section II: Free-Response Section = 50% of the exam grade

Students will have 80 minutes to answer 6 short-response questions and 2 long-response questions. The free-response section begins with a 10-minute mandatory reading period in which students can read the questions and plan their responses.

To achieve a high score on the free-response questions, students must provide ample scientific reasoning, relevant examples, and other appropriate evidence to support their answers.

AP Biology Exam Format at-a-Glance

SECTION I		
Question Type	**Number of Questions**	**Timing**
Part A: Multiple-Choice	63	90 minutes
Part B: Grid-In	6	

SECTION II		
Question Type	Number of Questions	Timing
Long Free-Response	2	80 minutes + 10-minute reading period
Short Free-Response	6	

Students are permitted to use a four-function calculator (with square root) to answer questions on both sections of the exam since both sections contain questions requiring data manipulation. To see which types of calculators are approved for the AP Biology exam, visit *http://www.collegeboard.org*.

As part of their testing packet, students will be given a list of formulas needed to answer quantitative questions that involve mathematical reasoning.

3. **Scoring the AP Biology Exam**

Total scores on the multiple-choice section of the exam are based on the number of questions answered correctly. Points are not deducted for incorrect answers or unanswered questions. The multiple-choice questions are scored by machine, and the free-response questions are scored by AP Exam Readers.

Free-response question scores are weighted and combined with the results of the computer-scored multiple-choice questions to obtain a raw score. This raw score is then converted into a composite AP score of 5, 4, 3, 2, or 1. These AP scores rate how qualified students are to receive college credit or placement:

AP Score Qualification
5: Extremely well qualified
4: Well qualified
3: Qualified
2: Possibly qualified
1: No recommendation

4. Using Materials to Supplement Your Crash Course

The AP Biology Topic Outline is published by the College Board as a guide for teachers to use in designing their AP Biology course. Every question on the AP exam can be directly tied back to one of the topics on the course outline. Sample multiple-choice and free-response questions can be found in the *revised* AP Biology Course Description. In addition, REA's *AP Biology All Access* Book + Web + Mobile study system further enhances your exam preparation by offering a comprehensive review book plus a suite of online assessments (topic-level quizzes, mini-tests, a full-length practice test, and e-flashcards), all designed to pinpoint your strengths and weaknesses and help focus your study for the exam. We strongly recommend that you use all of these materials to prepare for the new AP Biology exam.

5. Some Basic Test-Taking Strategies

One of the best ways to prepare for the AP Biology exam is to take the free online practice exam available with the purchase of this book. This practice exam will help you become familiar with the format of the new test and the types of questions you will be asked. The Detailed Explanations of Answers section will provide feedback that will help you to understand which questions give you the most difficulty. Then you can go back to the text of this book, reread the appropriate sections of your textbook, or ask your teacher for help on topics that still give you trouble. The more questions you answer in preparation for this test, the better you will do on the actual exam.

When you are studying for the AP Biology exam you do not need to study separately for the two sections of the test. As you prepare for the multiple-choice questions, you are also preparing for the free-response questions (FRQs). All of the questions relate back to the topics in the AP Biology topic outline.

On test day, remember to read all the questions carefully, and be alert for words such as *always, never, not*, and *except*. On the multiple-choice section, review all the answer choices before selecting your answer.

When preparing for the multiple-choice section, students often wonder if they should guess the answer to a question. Remember, there is no penalty for incorrect answers. Therefore, guessing is always advised if you have no idea of the correct answer. Before resorting to a blind guess, however, you should use all your knowledge and understanding of biology to eliminate the possible incorrect answers, so that any guess you are forced to make is an *educated guess*. Of course, you don't have to guess as there are also no points deducted from your score for unanswered questions.

When practicing for the test, give yourself enough time to answer all of the questions. The amount of time left in a given section will be announced by the proctor, but you must use your time wisely. Our online practice test with timed testing conditions will help you budget your time efficiently.

On the free-response section, make sure you write clearly. This sounds like a very simple thing, but if those who are scoring your exam cannot read your answer, you will not get credit. You should cross out any errors—using a single line through any mistakes—rather than erase them.

Also on the free-response section, pay particular attention to any questions that use the words *justify, explain, calculate, determine, derive*, and *plot*. All of these words have precise meanings. Pay attention to these words and answer the question as it's asked in order to receive maximum credit. Be sure to support your answer with examples and other scientific evidence. Avoid including irrelevant or extraneous material in your answer.

At this stage of your school career, it may be obvious to remind you of some basic preparations right before test day—but we will anyway because they're tried-and-true: get a good night's sleep the night before the exam, eat a good breakfast, and don't forget to bring a bunch of those famous No. 2 pencils.

6. What You Need for Exam Day

Here's a handy chart of what you bring with you to the exam and what you cannot have in the exam room:

Yes	No
• Several sharpened No. 2 pencils (with erasers) for completing the multiple-choice questions • One or two reliable dark blue or black ink pens for filling out the exam booklet covers and for answering the free-response questions. Avoid pens that clump or bleed. • A wristwatch, so you can monitor your time. Make sure it does not beep or have an alarm. • Your school's code if you are testing at a school different from the one you usually attend. • Your Social Security number (for identification purposes) • A government-issued or school-issued photo ID and your AP Student Pack if you do not attend the school where you are taking the exam • The College Board SSD Accommodations Letter if you are taking an exam with approved testing accommodations • Up to two calculators with the necessary capabilities for the AP Biology exam	• Cell phones, smartphones, tablet computers, MP3 players, and any electronic devices that can access the Internet • Cameras or other recording devices • Books, including dictionaries • Scratch paper • Mechanical pencils • Notes you've made in advance • Highlighters and colored pencils • Clothing with subject-related information • Food and drink

PART II
EVOLUTION

Natural Selection and Evolution

I. Natural Selection and Evolution

A. Contributing Ideas to Darwin's Descent with Modification

1. In the 1700s and 1800s, the biological sciences were defined in terms of *natural theology* rather than scientific data and extrapolation. Several scientists began to use data to debunk natural theology as a means for explaining scientific findings.

2. Charles Darwin built on the ideas of other scientists to develop his theory of "descent with modification" by natural selection.

3. While examining the fossil record, both Jean-Baptiste Lamarck and Charles Darwin agreed that species evolve over time, but each proposed a different mechanism:

   i. Lamarck proposed the (incorrect) mechanism for evolution—called the *inheritance of acquired characteristics*—which asserted that if a trait is used, it will be passed down to the next generation, but, if not used, then it will be discarded and not passed along. His theory is notable because of its emphasis on an organism's adaption to the environment.

   ii. Darwin also recognized that species change over time, but he proposed a *different mechanism* for how that change occurs, which he called natural selection.

4. Darwin was influenced by the ideas proposed by a number of other scientific thinkers, as well as his own extensive observations of biogeography and of plant and animal breeding.

     i. Charles Lyell, a geologist, proposed that the Earth had been around for a long period of time, that geological processes—such as volcanic eruptions—that occur presently also occurred in the past (*uniformitarianism*), and that these types of processes, over a long period of time, account for large-scale changes in the Earth's physical characteristics (*gradualism*).

    ii. These ideas led Darwin, and others, to conclude that the *strata* (and their fossils), observable in the exposed rock, represent distinct time periods during the Earth's history.

    iii. Thomas Malthus proposed that population size remains fairly steady, despite its capacity for exponential population growth because of disease, wars, and limited resources.

       ➤ Darwin felt that this situation applied more generally to all species and further proposed that the availability of limited resources led to competition between members of a species.

B. Natural Selection

  1. By studying 12 different types of finches on the Galapagos Islands, Darwin made a link between the origin of a new species and the environment in which these species reside.

  2. *Theory of Natural Selection*—reproductive success of an organism depends on its ability to adapt to the environment in which it resides. For example, several of the finches in the Galapagos Islands adapted their beak structure in order to find food.

  3. Postulates of Natural Selection—

    i. If the environment cannot support the individuals who occupy it, then competition occurs between members of a species and affects the production of offspring.

    ii. Survival of individuals within a population will depend on their genetic background. Individuals with traits that promote survival will pass these traits to offspring, allowing them to be more "fit" for their environment.

    iii. Over time, the fittest organism will survive, hence "survival of the fittest," and therefore changes in the population (genetic variation and mutations) cause variability and are an asset to a species. These population changes take place to benefit the reproduction of the population.

➤ The result of natural selection is the adaption of populations to their environment, thus giving them a competitive advantage to survive.

iv. A genetic variation, such as the average beak length of a finch that changes based on the season, is an example of adaption. Such a trait manifests itself in order to provide an advantage in specific environmental conditions. For example, during the dry season, the average beak length gets slightly larger, giving the finches a better advantage to traverse terrain and outcompete other birds for seeds that are less abundant in a wet season. A larger beak indicates a competitive advantage and survival of the fittest.

➤ Below is hypothetical data: dry seasons are 1950 and 1980; wet seasons are 1960, 1970, 1990, and 2000.

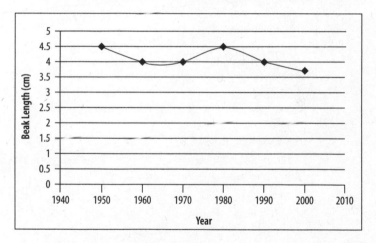

Darwinian biology permeates all aspects of biology. Early theories of evolution can make up a series of questions on the AP Biology exam. Be sure to familiarize yourself with the major evolutionary theories for the exam.

Evolution: An Ongoing Process

I. Evolution: An Ongoing Process

A. *Population Genetics*—study of genetic variation within a population of individuals.

1. *Population*—a group of individuals that belong to the same species.
2. *Gene Pool*—the total sum of genes within a population at a given time.

B. *Hardy-Weinberg Equilibrium*—study of the gene pool of a non-evolving population.

1. Hardy-Weinberg Equilibrium indicates that the frequencies of two alleles do not change from generation to generation; a population is said to be in Hardy-Weinberg Equilibrium if the following five conditions are met:
   i.   A very large population sample
   ii. No migration of individuals into or out of the population
   iii. No mutation
   iv. No natural selection
   v.   Random mating
2. To determine if a population is in Hardy-Weinberg Equilibrium, use this equation:
$$p^2 + 2pq + q^2 = 1$$
p = frequency of the dominant (homozygous) allele (A)
q = frequency of the recessive allele (a)
$2pq$ = frequency of dominant (heterozygous) allele (Aa)
   ➤ Keep in mind: the combined gene frequency must be 100% so that $p + q = 1$.

Sample Problem #1: Assume a population of 500 pea plants in which green is dominant to yellow. Use the chart below to see how to calculate the frequencies of all phenotypes.

A = green, a = yellow

Phenotype	Green	Green	Yellow
Genotype	AA	Aa	aa
Number of pea plants (total = 500)	320	160	20
Genotypic frequencies	320/500 = 0.64 AA	160/500 = 0.32 Aa	20/500 = 0.04 aa
Number of alleles in gene pool	320 x 2 = 640 A	160 A + 160 a = 320 A & a	20 x 2 = 40 aa 40 a
Allelic frequencies	640 A + 160A = 800 A 800/1000 = 0.8A p = frequency of A = 0.8		160a + 40aa = 200 a 200/1000 = 0.2 a q = frequency of a = 0.2

➤ $p^2 + 2pq + q^2 = 1$
- p^2 = frequency of AA = 0.8 × 0.8 = 0.64 = 64%
- $2pq$ = frequency of Aa = 2 × 0.8 × 0.2 = 0.32 = 32%
- q^2 = frequency of aa = 0.2 × 0.2 = 0.04 = 4%

➤ $p + q = 1$
- 0.8 + 0.2 = 1 (Always check to make sure these numbers = 1)

Sample Problem #2: Assume that in a population of insects, body color is being studied: 36% of the insects represent the orange color, which is recessive, and 64% represent the black dominant phenotype.

> If each successive generation maintains the allele frequency, the population is said to be in Hardy-Weinberg equilibrium.

1) Determine the allelic frequencies.
2) Determine the genotypic frequencies.
   i.   The recessive phenotype is key to this problem because the dominant represents both AA and Aa. However, recessive is *only* represented by aa. Use logic that q^2 = aa; therefore, the square root of .36 or q = 0.6. Since $p + q = 1$, $p + 0.6 = 1$, then $p = 0.4$.
   ii. Allelic frequencies are A = 0.4, a = 0.6
   iii. Genotypic frequencies follow the equation
        $$p^2 + 2pq + q^2 = 1$$
   iv. $p^2 = (0.4)^2 = 0.16 = 16\%$
        (AA or homozygous dominant) → Black phenotype
        $2pq = 2 \times 0.6 \times 0.4 = 0.48 = 48\%$
        (Aa, heterozygous dominant) → Black phenotype
        $q^2 = (0.6)^2 = 0.36 = 36\%$
        (aa, homozygous recessive) → Orange phenotype
        16 + 48 + 36 = 100 (Always double-check your numbers)

Test Tip

Know the Hardy-Weinberg Equilibrium concept and how to use it. It is highly likely that there will be questions on the AP Biology exam that refer to it.

C. *Microevolution*—the change in the frequencies of alleles or genotypes in a population from generation to generation (evolution on a small scale) occurs if any of the five conditions of Hardy-Weinberg equilibrium are *not* met.

1. *Genetic Drift*—defined as changes in the gene pool due to chance because of a small population. The small population directly contrasts the large population needed to maintain Hardy-Weinberg equilibrium.

      i. Causes a significant, genetic change (microevolution) of a species if only a few members of a population migrate to found a new population.

     ii. Causes genetic change (microevolution) anytime a species is reduced to very small numbers due to chance events, such as hurricanes, earthquakes, fires, or habitat destruction.

2. *Bottleneck Effect*—changes in the gene pool due to some type of disaster or massive hunting that inhibits a portion of the population from reproducing. The small population directly contrasts with the large population needed to maintain Hardy-Weinberg equilibrium.

3. *Founder Effect*—a new colony is formed by a few members of a population, so the smaller the sample size, the less the genetic makeup of the population. The small population directly contrasts with the large population needed to maintain Hardy-Weinberg equilibrium.

4. *Gene Flow*—transfer of alleles from one population to another through migration. The gametes of fertile offspring mix within a population, providing genetic variation. Genetic variation directly contrasts the no gene-flow postulate needed to maintain Hardy-Weinberg equilibrium.

5. *Mutation*—a change in the genetic makeup of an organism at the DNA level. Mutation directly contrasts the no mutation postulate needed to maintain Hardy-Weinberg equilibrium.

6. *Non-random Mating*—individuals mating with those in close vicinity. Non-random mating directly contrasts with the random mating postulate needed to maintain Hardy-Weinberg equilibrium.

7. *Natural Selection*—reproductive success of organisms depends on their ability to adapt to the environment in which they reside. Natural selection directly contrasts with the no natural selection postulate needed to maintain Hardy-Weinberg equilibrium.

   ➤ Keep in mind that any *genetic variation within a population can increase that population's genetic diversity*, even within the same species.

➤ Some *phenotypic variations can significantly increase or decrease the fitness of an organism* and the overall population.

    Examples include DDT resistance in insects, the peppered moth, and Sickle cell anemia.

➤ *Humans can also impact other species* through: loss of genetic diversity within a crop species, overuse of antibiotics, and artificial selection.

D. *Speciation*—the origin of new species (a population of individuals who can mate with each other and produce viable offspring).

1. How Does It Occur?

    i. *Allopatric Speciation*—populations are separated by geographical isolation, thus a new species can be formed following adaption to new surroundings.

    ii. *Adaptive Radiation*—Evolution of a large number of species from a common ancestor. The finches Darwin found on the Galapagos Islands are an example of adaptive radiation.

    iii. *Sympatric Speciation*—populations are not separated by geographical isolation, but a new species is formed within the parent populations.

        ➤ *Autopolyploidy*—meiotic error causes a species to have more than two sets of chromosomes. Contribution is from one species.

        ➤ *Allopolyploidy*—polyploidy is a result of two different species.

2. How Fast Does It Occur?

    i. *Gradualism*—species are produced by slow evolution of intermediate species.

    ii. *Punctuated Equilibrium*—speciation occurs quickly at first and then is followed by small changes over a long period of time.

Test Tip

Natural selection acting on a population is the mechanism by which a species' characteristics change (evolve) over time. Remember to think about how one topic in biology relates to another.

E. *Modes of Natural Selection*—natural selection will favor the allelic frequency in three ways. Below is an example of a bell curve or normal distribution population. Three types of selections can take place that shift the bell curve to different frequencies (hence, evolution is taking place).

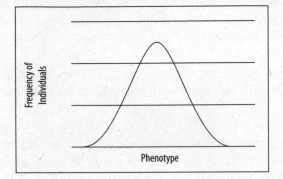

Original frequency of individuals shows a normal "bell-curve" distribution

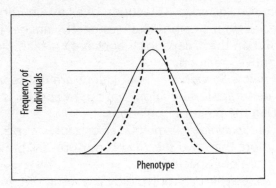

Stabilizing Selection—Extreme phenotypes are removed and more common phenotypes are selected

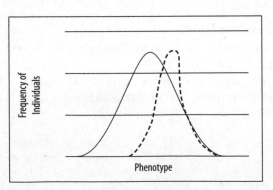

Directional Selection—One of the extreme phenotypes is selected

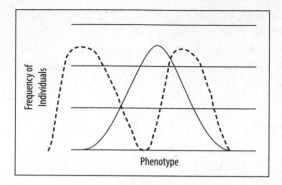

Diversifying Selection—Both of the extreme phenotypes are selected

Do you like math? Let's hope so because there may be easy mathematical calculations on the AP Biology test. For example, you should know how to appropriately use the Hardy-Weinberg equation: $p + q = 1$, $p^2 + 2pq + q^2 = 1$. Also make sure you understand how to use the chi-square equation—don't memorize it. If needed, it will be given to you: $X^2 = \Sigma \left[\dfrac{(o - \theta)^2}{\theta} \right]$.

Common Descent

I. Common Descent

A. Evidence for Evolution

1. *Biogeography*—study of organisms and how they relate to the environment. Some organisms may be unique to certain geographies; hence, those organisms have adapted to live in that environment.

2. Fossils—help indicate the progression of organisms from simple to complex. For example, transitional fossils are fossils of animals that display a trait that helped the organism attain a competitive advantage. At one time, whales had limb-like appendages indicating they may have been land dwellers.

3. *Comparative Anatomy*—study of anatomical similarities between organisms.
   i. *Homologous structures*—structures in organisms that indicate a common ancestor. For example, a human arm, cat leg, whale flipper, and bat wing all have a similar structure, but different functions.
   ii. *Vestigial organs*—remnants of structures that were at one time important for ancestral organisms.

4. *Comparative Embryology*—comparing the embryonic development of one organism to another.

5. *Molecular Biology*—used in the study of evolution by looking at homology in DNA and protein sequences and genes; this study allows for an even broader level of comparison between organisms as different as prokaryotes, plants, and humans.

➤ Organisms share conserved core processes, which signal their evolution from a common ancestor and how widely distributed these processes have become among different species

➤ *Examples:* DNA and RNA are carriers of genetic information through transcription, translation, and replication; the genetic code of many organisms is shared and is evident in many modern living systems; and many metabolic pathways, like glycolysis, are conserved.

➤ Structural evidence, such as cytoskeletons, membrane-bound organelles, linear chromosomes, and endomembrane systems, suggest that all eukaryotes are related.

B. Evolution Continues to Occur

1. Scientific evidence supports the premise that evolution continues to occur.
   Examples include:
   ➤ Emergent diseases
   ➤ Chemical resistances caused by mutations, such as resistance to antibiotics, pesticides, herbicides, and chemotherapy drugs
   ➤ Phenotypic change in a population (such as Darwin's finches in the Galapagos)
   ➤ Eukaryotes eventual development of structures such as limbs, brain, and immune system

C. Phylogenic Trees and Cladograms

1. Represent traits that are either derived or lost due to evolution, such as opposable thumbs, the absence of legs in some sea animals, and the number of heart chambers in animals.
2. Illustrate that speciation has occurred and when two groups were derived from a common ancestor.
3. Can be constructed from either morphological similarities or from DNA and protein sequence similarities by utilizing a computer program that measures the organisms' interrelatedness.

4. Provide a dynamic snapshot that is constantly being revised.
Example

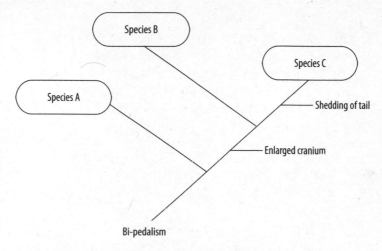

➤ Species B and C are more closely related to each other than to species A.
➤ All species are generated from an ancestor species with bi-pedalism.
➤ All species retain traits from the ancestor but have evolved to gain some specific trait through time.

Organism	Bi-pedal	Large Cranium	Tail Loss
Species A	X		
Species B	X	X	
Species C	X	X	X

Origin of Life

I. Origin of Life

A. The Origin of Life: Hypotheses and Evidence

1. Primitive Earth was thought to have the following atmospheric molecules — water (H_2O), methane (CH_4), hydrogen (H_2), and ammonia (NH_3) — and no oxygen.

2. These inorganic precursors of organic molecules on primitive Earth could have been formed as a result of an electrical spark and the lack of oxygen.

3. As a result, crude organic molecules including sugars, lipids, amino acids, and nucleic acids were formed.
   i. *Miller-Urey Experiment* — tested the Oparin-Haldane model; the atmosphere on primitive Earth was the precursor for the synthesis of organic molecules.
   ii. *Heterotrophic Hypothesis* — first forms of life were prokaryotic heterotrophs that produced organic matter.

4. Molecules then became the building blocks of more complex molecules. (Examples: amino acids and nucleotides.)

5. Monomers then began joining to form polymers that, over time, began to replicate, store, and transfer information.

6. Complex reactions could have occurred in a solution, known as the *Organic Soup Model*, or as reactions on solid reactive surfaces.

7. *RNA (ribonucleic acid)* — the first genetic material; it is capable of self-replication and can act as both genotype and phenotype. Eventually, *DNA (deoxyribonucleic acid)* became the genetic material because of its stability over RNA and its ability to correct mutations.

B. Earth's History

1. Geographical Evidence:
   i.   Earth is most likely around 5 billion years old.
   ii. Earth's environment was too hostile for life until about 3.9 billion years ago.
   iii. Earliest fossil records date back 3.5 billion years ago.

2. Molecular and Genetic Evidence:
   i.   *Anaerobic prokaryotes* emerged approximately 4 billion years ago and represent the first origins of life.
   ii. Earliest living organisms were unicellular, had a genetic code, and were able to evolve and reproduce.
   iii. Prokaryotes diverged into two types—bacteria and archaea—about 2.5 billion years ago.
   iv. Oxygen accumulated in the atmosphere approximately 2.5 million years ago as a result of photosynthetic bacteria.
   v.   Eukaryotes emerged 2 billion years ago via the *Endosymbiotic Theory*.
   vi. Prior to 500 million years ago, life was confined to aquatic environments. Plants eventually found a foothold on earth (root system) via a symbiotic relationship with fungi.
   vii. All living things come from a common ancestor.

C. Extinction and Adaptive Radiation

1. Extinction of a species is very common, and more than a dozen mass extinctions have occurred throughout geological history.
   i.   Extinction rates become rapid during ecological stress.
   ii. For example, during the *Cretaceous extinction*, which occurred approximately 65 million years ago, about 50% of species, including almost all of the dinosaurs, became extinct.

The names and dates of the major extinctions will not be on the exam; however, be prepared to use data to determine that extinction has occurred.

2. Adaptive radiation—the rapid development of new species from a common ancestor; may occur after a significant genetic change in a member of a species, or after a new habitat becomes available due to extinction of another (or many) species.
   i.   Adaptive radiation causes an increase in speciation.
   ii. Occurs after mass extinctions.
   iii. Some significant adaptive radiations include the radiation of flowering plants after the development of effective dormancy and dispersal strategies (e.g., pollen and seeds) and the adaptive radiation of mammals after the mass extinction of dinosaurs.

PART III

CELLULAR PROCESSES:
Energy and Communication

Energy

I. Chemistry of Life—Energy Changes

A. *Energy*—capacity to do work.

　1. *1st Law of Thermodynamics*—energy can neither be created nor destroyed but can change from one form to another and be transferred.
　　Example: Plants convert light energy from the sun to make glucose, a form of chemical energy.
　2. *2nd Law of Thermodynamics*—every energy transfer increases entropy of the universe (disorder).
　　i. All living systems will not violate the 2nd Law of Thermodynamics.

B. *Free Energy*—energy available in a system to do work; organisms need this free energy in order to maintain organization, to grow, and to reproduce.

　1. *Exergonic* reactions release free energy.

$$AB \rightarrow A + B + Energy$$

　　i. In *catabolic reactions*, reactant(s) are broken down to produce product(s) containing less energy.
　　ii. The energy released can be used for reactions that require energy.
　2. *Endergonic* reactions require free energy.

$$A + B + Energy \rightarrow AB$$

　　i. In *anabolic reactions*, reactant(s) are joined together to produce product(s) containing more energy.

      ii. The free energy required by anabolic reactions is often provided by ATP produced in catabolic reactions.

  3. *Adenosine triphosphate (ATP)* carries energy in its high energy phosphate bonds.

      i. ATP is formed from adenosine diphosphate (ADP) and inorganic phosphate.

$$ADP + Pi + Energy \rightarrow ATP$$

      ii. Conversely, when ATP is broken down into ADP and Pi via hydrolysis, energy is released (*exergonic*) that can be used in *endergonic* reactions.

      iii. In addition, ATP can donate one of its phosphate groups to a molecule, such as a substrate or a protein, to energize it or cause it to change its shape.

  4. Living systems require a consistent input of free energy and an ordered system.

      i. This free energy input allows for a system's order to be maintained.

      ii. If either order in the system or free energy flow were to occur, death can result.

      iii. Biological processes are in place to help offset increased disorder and entropy and to help maintain order within a system; therefore, energy input into the system must exceed the loss of free energy in order to maintain order and to power cellular processes.

      iv. Energy storage and growth can result from excess acquired free energy beyond the required energy necessary for maintenance and order within a system.

      v. Changes in free energy can affect population size and cause disruptions to an ecosystem.

  5. *Metabolism*—the totality of all chemical reactions that occur within an organism.

      i. Reproduction and rearing of offspring require free energy beyond what is normally required for the maintenance and growth of the organism. Energy availability can vary, and different organisms utilize a variety of reproductive strategies as a consequence. Some examples include seasonal reproduction by animals and plants and life-history strategy (biennial plants, reproductive diapause).

ii. Organisms utilize free energy in order to help regulate body temperature and metabolism. Mechanisms through which these are done include:
> *Endothermy*—use of internal thermal energy that is generated by metabolism to maintain an organism's body temperature.
> *Ectothermy*—use of external thermal energy to assist in the regulation of an organism's body temperature.
> Some plant species utilize elevated floral temperatures.

iii. There is an important relationship between the metabolic rate/unit body mass and the size of multicellular organisms. In other words, smaller organisms generally have higher metabolic rates.

C. Energy Coupling

1. *Coupled reactions*—a chemical reaction having a common intermediate in which energy is transferred from one reaction to another.

2. A system can maintain order by utilizing coupling cellular processes that increase entropy (causing negative changes in free energy) with those that decrease entropy (causing positive changes in free energy).

3. The molecule that is essential for coupling reactions and cellular work is ATP.

4. Exergonic reactions, like ATP $\rightarrow$ ADP, is an example of an energetically favorable reaction because it allows for a negative change in free energy that will then be used in order to maintain or to increase order within a system that is coupled by reactions that demonstrate changes in positive free energy.

5. The processes of cellular respiration and photosynthesis are coupled to each other. The products of one reaction end up being the reactants in the other.

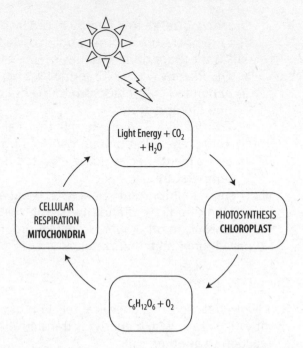

6. Electron transport and oxidative phosphorylation are examples of coupled reactions.

D. Modes of Energy Capture

1. Organisms can capture and store free energy for nutritional use in their biological systems.
    i. *Autotrophs*—"self-feeders," create their own organic molecules or food; they are known as producers.
    ii. *Heterotrophs*—cannot create their own organic molecules or food; they are known as consumers.
        ➤ *Hydrolysis*—helps them metabolize carbohydrates, proteins, and lipids as sources of free energy.

➤ The following chart shows modes of nutrition:

Mode of Nutrition	Description; Examples (Other Nonprokaryote Examples)
Photoautotrophy	Use light as an energy source and gain carbon from CO_2; cyanobacteria (also plants and some protists)
Chemoautotrophy	Use an inorganic energy source and gain carbon from CO_2; some archaebacteria
Photoheterotrophy	Use light as an energy source and gain carbon from organic sources; some prokaryotes
Chemoheterotrophy	Use an organic energy source and gain carbon from organic sources; most prokaryotes (also animals, fungi, and some protists)

2. Biological systems can capture energy at multiple points in their energy-related pathways. Some examples of these pathways include the *Krebs cycle*, *glycolysis*, the *Calvin cycle*, and *fermentation*.

3. Energy capturing processes, such as NADP+ in photosynthesis and oxygen in cellular respiration use different types of electron acceptors.

Note: For more on photosynthesis, see Chapter 7; for more on cellular respiration, see Chapter 8.

Names of enzymes and specific steps and intermediates of pathways—The exam will not require you to know these details; however, be prepared to understand the concepts in this chapter and how organisms utilize free energy.

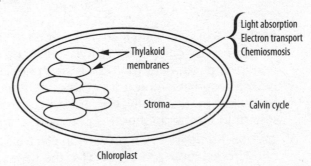

Photosynthesis

I. Key Concepts

A. Photosynthesis occurs in all photosynthetic autotrophs, including plants, algae, and photosynthetic prokaryotes.

B. In eukaryotes, photosynthesis occurs in chloroplasts; in prokaryotes, it occurs in the plasma membrane and in the cytoplasm.

C. The overall equation for photosynthesis is:

$$6CO_2 + 6H_2O + \text{light energy} \rightarrow C_6H_{12}O_6 + 6O_2$$

D. Photosynthesis is affected by a variety of environmental factors.

E. In eukaryotes, each phase of photosynthesis takes place in the chloroplasts.

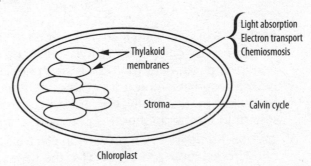

Light absorption
Electron transport
Chemiosmosis

Thylakoid membranes

Stroma —— Calvin cycle

Chloroplast

The Two Steps of Photosynthesis

A. Photosynthesis has two main steps:

1. *Light-dependent reactions*—the absorption of light energy and its conversion to the chemical energy of ATP and the reducing power of NADPH.
2. *Light-independent reactions*—the use of ATP and NADPH to convert CO_2 to sugars using the Calvin Cycle.

B. STEP 1—*Light-dependent reactions* occur in the *thylakoid* membranes of chloroplasts in eukaryotes.

1. Pigment molecules collect light energy.
   i. *Chlorophyll a*—main photosynthetic pigment
   ii. *Chlorophyll b* and *carotenoids*—accessory pigments that allow leaves to capture a wider spectrum of visible light than chlorophyll a alone
   iii. The following graph shows the absorption spectra of photosynthetic pigments:

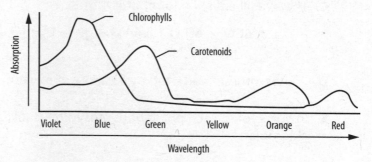

2. *Photosystems (PS) I and II* are embedded in the internal membranes of chloroplasts (thylakoids) and consist of hundreds of pigment molecules that funnel light energy to two chlorophyll a molecules at the reaction center of each photosystem. Essentially, they use an electron transport system to transfer higher free energy electrons.
3. *Electron Transport Chain (ETC)*—an electrochemical gradient of hydrogen ions (protons) across the thylakoid membranes that undergoes redox reactions in a series.

   i. Electrons are transferred from PSII → primary electron acceptor → until donated to PSI → next electron carriers → donated to $NADP^+$ to reduce it to NADPH.

   ii. Electrons can take either a non-cyclical route or a cyclical one. The primary difference between the two is that the cyclical flow of electrons produces more ATP and takes place because the Calvin Cycle uses more ATP per mole than NADPH per mole, and hence replenishes the used ATP.

   iii. The proton gradient is linked to the synthesis of ATP and ADP and inorganic phosphate via ATP synthase.

4. *Chemiosmosis*—the movement of H^+ ions down their concentration gradient from inside the thylakoids to the stroma. As they do this, they pass through the enzyme, *ATP synthase*, which causes the catalysis of ATP from ADP and Pi.

   i. The following figure shows both the electron transport and chemiosmosis of photosynthesis.

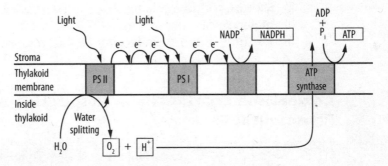

5. *Flow summary*—absorption and conversion of light energy to ATP and NADPH

   i. Pigments absorb light energy.

   ii. Light energy sends electrons down the electron transport chain.

   iii. The electrons eventually reduce $NADP^+$ to NADPH.

   iv. Water is split, forming e^-, H^+, and O_2.

   v. H^+ concentration builds up inside the thylakoids (the thylakoids space).

   vi. When H^+ move through ATP synthase from the thylakoid space to the stroma, ATP is formed.

   vii. NADPH and ATP are used in the second step of photosynthesis, carbon fixation.

C. STEP 2—*Calvin Cycle/Light Independent Cycle*

1. Occurs in the stroma of chloroplasts
2. Uses the products (ATP and NADPH) to produce glucose
3. The following figure depicts the steps in the *Calvin Cycle*:

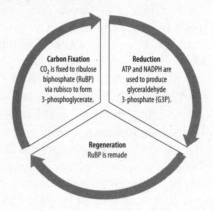

Carbon Fixation
CO_2 is fixed to ribulose biphosphate (RuBP) via rubisco to form 3-phosphoglycerate.

Reduction
ATP and NADPH are used to produce glyceraldehyde 3-phosphate (G3P).

Regeneration
RuBP is remade

i. Six turns of the cycle fix six carbons, representing one molecule of glucose.
ii. Six turns of the cycle require 18 ATP and 12 NADPH.

III. Comparison Chart: Cellular Respiration and Photosynthesis

Process	Takes Place in Cellular Respiration	Takes Place in Photosynthesis
Breakdown of glucose	Yes	No
Synthesis of glucose	No	Yes—Calvin Cycle
O_2 is released	No	Yes—light-dependent reaction
O_2 is consumed	Yes—ETC and oxidative phosphorylation	No

(continued)

Comparison Chart: Cellular Respiration and Photosynthesis (*continued*)

Process	Takes Place in Cellular Respiration	Takes Place in Photosynthesis
Chemiosmosis	Yes—ETC	Yes—ETC
CO_2 is released	Yes—Shuttle Step and Krebs Cycle	No
CO_2 is consumed/ fixed	No	Yes—Calvin Cycle
ATP is produced	Yes—glycolysis, Krebs Cycle, ETC, and Oxidative Phosphorylation	Yes—light-dependent reaction
ATP is consumed	Yes—glycolysis initial investment	Yes—Calvin Cycle
Pyruvate as intermediate	Yes—glycolysis	No
NADH produced	Yes—glycolysis, shuttle step, Krebs Cycle	No
NADPH produced	No	Yes—light-dependent reaction

Test Tip

Familiarizing yourself with the similarities between cellular respiration and photosynthesis is recommended. It's been a popular test item on past exams. Be sure you know about ATP production, electron transport use, compartmentalization between chloroplast and mitochondria, hydrogen and electron acceptor molecules, such as NADH, FADH$_2$, and NADPH.

Fermentation and Cellular Respiration

Key Concepts

A. Cellular respiration is the catalysis (breakdown) of glucose to produce energy (ATP) and organic intermediates used in the synthesis of the other organic molecules (amino acids, lipids, etc.) needed by the cell.

B. Some form of cellular respiration takes place in nearly all organisms.

   1. *Glycolysis* is the oldest metabolic pathway, is virtually universal, and takes place in the cytoplasm of cells.
   2. Aerobic respiration—the Krebs Cycle, electron transport, and chemiosmosis—takes place in mitochondria in eukaryotes.

C. Refer to this overall equation for cell respiration:

$$C_6H_{12}O_6 + 6O_2 \rightarrow 6CO_2 + 6H_2O + Energy$$

Although this equation is almost the reverse of the equation for photosynthesis, the two processes involve different enzymes and biochemical pathways, as well as different organelles.

D. Cells may utilize an anaerobic pathway (fermentation) that does not require O_2, or an aerobic pathway that does require O_2.

   1. Glycolysis is the first step of both pathways. This step does not require O_2.

2. Aerobic respiration has three additional steps, the second of which requires O_2, as the final electron receptor of the electron transport chain.
   i. The Krebs Cycle takes place in the matrix of the mitochondria.
   ii. The electron transport chain takes place in the inner membrane of the mitochondria.
   iii. Chemiosmosis takes place across the inner membrane of the mitochondria.
   iv. The following figure shows the location of fermentation and the steps of cellular respiration in the mitochondrion:

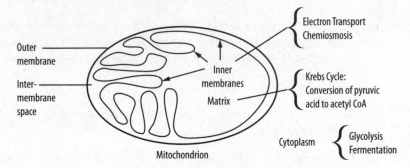

3. Anaerobic fermentation has one additional step following glycolysis that regenerates the oxidizing agent, NAD^+, to allow glycolysis to operate in the absence of O_2.
4. Both fermentation and cellular respiration are catabolic and involve oxidation-reduction reactions:
   i. Loss of electrons is oxidation (glucose to carbon dioxide).
   ii. Gain of electrons is reduction (oxygen to water).
   iii. Electrons = Energy

Test Tip

Anaerobic fermentation produces much less energy than aerobic respiration: only 2 ATP per glucose processed, as compared to 36 ATP produced by aerobic respiration per glucose molecule.

II. The Four Parts of Cellular Respiration

A. *Glycolysis*—a ten-step metabolic pathway, catalyzed by a series of enzymes, which breaks one glucose molecule down to two molecules of pyruvic acid (pyruvate).

  1. The following are the most important features to remember about glycolysis reaction series:
     i.   The bonds of the glucose molecule are rearranged.
     ii. NAD^+ is reduced to NADH, one of the two electron carriers in cellular respiration.
     iii. Free energy is then released in the form of ATP—which comes from ADP and inorganic phosphates.
     iv. Pyruvic acid is produced and is transported from the cytoplasm to the mitochondria for future oxidation.
     v.   The following figure shows the process of glycolysis and its end products:

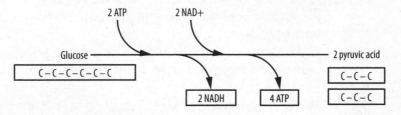

B. *Shuttle Step*—the conversion of pyruvic acid to acetyl-CoA occurs in the matrix of the mitochondria and involves three important features:

  1. Coenzyme A (CoA) is added.
  2. Pyruvate is oxidized, producing NADH.
  3. The 3-carbon pyruvate is converted to the 2-carbon acetyl CoA, releasing a molecule of CO_2.

4. The following figure shows the conversion of pyruvic acid to acetyl CoA:

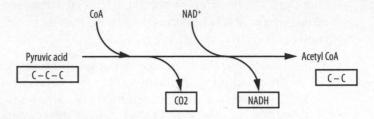

C. *Krebs Cycle*—series of reactions that continually regenerates one of its first reactants, *oxaloacetic acid* and contains the following important features:

1. The Krebs Cycle produces the majority of NADH, $FADH_2$, and CO_2 (waste product) for cellular respiration.
2. NADH and $FADH_2$ carry electrons that were extracted during the Krebs Cycle reactions and carries them to the electron transport chain.
3. ATP is synthesized from ADP and inorganic phosphates via phosphorylation, and coenzymes capture electrons during this cycle.
4. The key intermediate, *oxaloacetate (OAA)* is added to acetyl CoA to make *citrate*, which starts the entire Krebs Cycle.

D. *Electron Transport Chain*—carried out by electron carriers that undergo a series of redox reactions as electrons are passed from one carrier to another.

1. The following are the important features to know about electron transport:
   i. Occurs in the mitochondria and is similar to the ETC of photosynthesis.
   ii. NADH and $FADH_2$ deliver electrons and pass them to a series of electron acceptors as these electrons move toward the final electron acceptor, O_2.
   iii. This passage of electrons within the mitochondria is accompanied by a proton gradient that facilitates the electrons' movement down the ETC.

    iv. In the ETC, no direct ATP is made; it must be coupled to *oxidative phosphorylation* via chemiosmosis (or the diffusion of H^+ ions across the membrane).

      ➤ Cellular respiration produces a total of 38 ATP.

III. Fermentation

A. Glycolysis is the first step of both aerobic respiration and fermentation.

1. Two of the end products of glycolysis, pyruvic acid and NADH, can be processed anaerobically in the cytoplasm of certain cells.
2. The second step of fermentation does not produce ATP directly.
3. Rather, it generates NAD^+, which is required to keep glycolysis running and producing ATP.

B. Lactic acid fermentation includes glycolysis plus an additional reaction that generates NAD^+ and lactic acid.

1. Certain fungi, bacteria, and muscle cells have special enzymes that carry out lactic acid fermentation.
2. In vigorously exercising muscle cells, lactic acid fermentation provides ATP when the circulatory system cannot keep up with the oxygen demands of the muscle cells.

C. Alcohol fermentation includes glycolysis plus additional reactions that produce NAD^+, ethanol, and CO_2.

1. Single-celled organisms, such as yeast and some plant cells, have special enzymes to carry out alcohol fermentation.
2. Yeast is used in bread making because CO_2 gas causes bread to rise; the ethanol is removed by subsequent baking. Yeast is also used in beer making because it produces ethanol; CO_2 in an enclosed container produces carbonation.

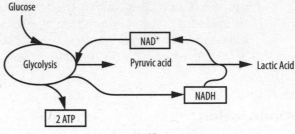

Lactic Acid Fermentation

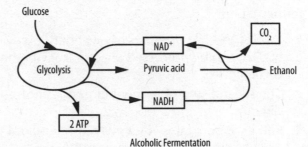

Alcoholic Fermentation

There is almost always at least one test question that requires comparing photosynthesis and cellular respiration. Also, keep in mind that plants carry out both photosynthesis and cellular respiration. Be sure to review the photosynthesis and cellular respiration chart found in Chapter 7: Photosynthesis.

Matter

I. Key Concepts

A. Because they are composed of matter, the basic rules of chemistry apply to all living organisms.

B. All organisms require an input of energy from the environment, as well as the means to control the orderly use of that energy.

1. Organisms generally convert the energy they obtain into ATP, the cell's "energy currency," which they use to power all life processes, including biosynthesis.
2. Biochemical reactions, catalyzed by a large array of enzymes that are specific for each reaction, control biosynthesis—the chemical reactions that produce the macromolecules of which an organism's cells are composed.

C. Biological molecules include carbohydrates, lipids, proteins, and nucleic acids that have a variety of important functions.

D. Water is the most abundant molecule in living organisms and possesses a variety of instrumental properties that result from its hydrogen bonding.

E. Enzymes are proteins that act as catalysts to speed up biochemical reactions.

II. **Water**

A. Water has many significant characteristics important to living organisms, primarily because of its polarity and hydrogen bonding.

B. As an *aqueous solvent*, many biochemical reactions can take place within a cell and in its immediate environment (the space between cells in a multicellular organism).

C. The pH of the aqueous environment inside a cell and its organelles influence many biological activities such as the shape of proteins, the creation of proton gradients across membranes, and the speed at which enzymes catalyze reactions.
   1. Acidic solution—contains more H^+ than OH^-
   2. Basic solution—contains more OH^- than H^+

D. The following are the properties of water that make it an important molecule:
   1. *Cohesion*—ability of water molecules to stick together
   2. *Adhesion*—ability of water to adhere to other molecules
   3. *Heat of Fusion (Surface Tension)*—difficulty in breaking the surface of water
   4. *High Specific Heat Capacity and Thermal Conductivity*—it heats up and cools down slowly
   5. *Heat of Vaporization*—water's high specific heat prevents it from evaporating easily
   6. *Universal Solvent*—many reactions can take place in water

 III. **Monomers and Polymers**

A. *Monomers* are building blocks of larger macromolecules called polymers.

B. *Macromolecules* are large molecules that fall into four categories: carbohydrates, lipids, proteins, and nucleic acids.

C. *Condensation reactions* are responsible for the biosynthesis of polymers from monomers with the removal of water. The following figure shows the condensation synthesis of a polymer:

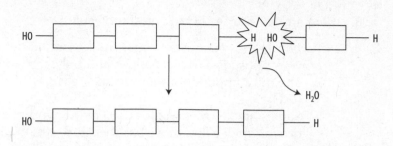

D. Conversely, *hydrolysis reactions* break down polymers into their monomers with the addition of water, and it is the reverse of the reaction shown in the figure on condensation synthesis of a polymer.

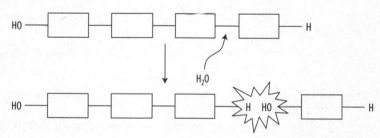

IV. Biological Molecules

A. These biologically important elements are found in all organisms:

1. *Carbon* (C), *hydrogen* (H), and *oxygen* (O) are found in all macromolecules.
2. Additionally, *nitrogen* (N) is found in significant amounts in proteins and nucleic acids.
3. *Sulfur* (S) is an element commonly found in proteins.
4. *Phosphorous* (P) is prominent in nucleic acids.

5. *Sodium* (Na), *potassium* (K), *magnesium* (Mg), and *iron* (Fe) are examples of important elements found in lesser quantities in most organisms.

B. There are four classes of biological molecules:

1. *Carbohydrates*—consist of sugar (monosaccharides) and polymers of sugars (disaccharides and polysaccharides).
   i. The most important monosaccharide is glucose ($C_6H_{12}O_6$).
   ii. Sugars are key metabolites used in the synthesis of other organic molecules, as well as the substrates of glycolysis and the products of photosynthesis.
   iii. The particular bonding between carbohydrate subunits is what determines the specific orientation of the carbohydrate and its secondary structure.

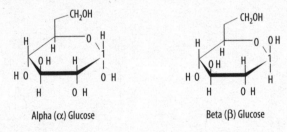

Alpha (α) Glucose   Beta (β) Glucose

2. *Lipids* — water-insoluble molecules composed of glycerol and fatty acids.
   i. *Fats* (triglycerides) are energy storage molecules consisting of one glycerol molecule with three fatty acid molecules attached.
      ➤ *Saturated fatty acids*—do not contain a double bond and are more likely to be a solid at room temperature.

      ➤ *Unsaturated fatty acids*—have one or more double bonds and are more likely to be fluid at room temperature.

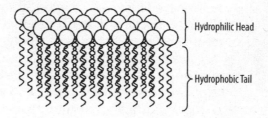

ii. *Phospholipids*—consist of one glycerol molecule with two fatty acid molecules attached as well as a polar component; they can self-assemble into a classic bilayer arrangement that is the basis of all biological membranes.

Hydrophilic Head

Hydrophobic Tail

iii. *Wax and steroids*—(including cholesterol) lipids with more complex structures that have a variety of functions. The following is the structure of the precursor lipid cholesterol:

H_3C CH_3

H_3C CH_3

CH_3

CH_3

HO

3. *Proteins*—polymers made up of different combinations of 20 commonly occurring amino acid monomers.
   i. Proteins have a wide variety of functions, including structural components of cells and tissues, transport proteins in the cell's membranes, and as catalysts called enzymes.
   ii. Amino acids share the same basic structure:

$$H_2N - \underset{\underset{R}{|}}{\overset{\overset{H}{|}}{C}} - \overset{\overset{O}{\|}}{C} - OH$$

➤ Connected by a linear sequence through the formation of peptide bonds by dehydration synthesis

➤ Contain a central carbon atom covalently bonded to four atoms or functional groups:

— One of the four is always a hydrogen atom

— A carboxyl functional group (acidic) –COOH and an amine functional group (basic) –NH_2

— Fourth component is a variable R group, which is different for each amino acid

iii. Proteins have four levels of physical structure:

➤ *Primary structure*—refers to the specific linear sequence of amino acids in a polypeptide.

➤ *Secondary structure*—the initial folding patterns of certain lengths of the polypeptide chain, such as alpha helices and beta pleated sheets.

➤ *Tertiary structure*—refers to the overall shape in which a polypeptide eventually folds.

➤ *Quaternary structure*—arises from the association of two or more folded polypeptides to form a multi-subunit protein.

4. *Nucleic Acids (DNA and RNA)*—made from monomers called nucleotides.

i. A nucleotide has three parts:

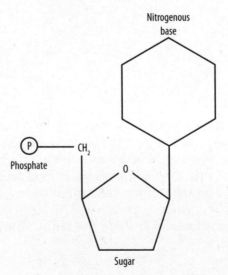

Nitrogenous base

Phosphate

CH_2

O

Sugar

➤ A *5 carbon sugar*—either deoxyribose (in DNA) or ribose (in RNA).

➤ *A phosphate group*—present on a nucleotide.

➤ One of the four *nitrogenous bases* is present in each nucleotide.

— Adenine, thymine, cytosine, and guanine (in DNA)

— Adenine, uracil, cytosine, and guanine (in RNA)

ii. Nucleic acids have ends, defined by the 3′ and 5′ carbons of the nucleotide's sugar. The direction in which other nucleotides are added to the chain during DNA synthesis and the direction in which transcription occurs are both determined by the placement of these two nucleotide ends.

C. Types, Functions, and Examples of Biological Molecules

Type of Biological Molecule	Examples	General Functions
Carbohydrates	Monosaccharides (sugars)	
	Glucose	Energy; building blocks of other carbohydrates
	Deoxyribose and ribose	DNA and RNA
	Polysaccharides	
	Starch and glycogen	Energy storage
	Cellulose	Plant cell wall structure
Lipids	Fats	Energy storage
	Phospholipids	Plasma membrane structure
	Waxes	
	Steroids (cholesterol)	Physical protection
		Hormones (part of cell membranes)

(continued)

Types, Functions, and Examples of Biological Molecules (*continued*)

Type of Biological Molecule	Examples	General Functions
Proteins	Enzymes Other proteins	Biochemical catalysts Structure, movement, signal reception, etc.
Nucleic Acids	DNA RNA ATP	Storage of genetic information Converts genetic information into proteins Energy currency of the cell

➤ Keep in mind that variations within these biological molecules allow for cells and organisms to possess a much wider variety of functions, such as having different types of hemoglobin or different phospholipids on a cell membrane.

Test Tip

Biologically important molecules are great examples of the "structure and function" theme. Be sure to use these examples in essay questions to gain critical points.

D. *Enzymes*—proteins that act as catalysts to speed up biochemical reactions.

1. The function of enzymes is to lower the *activation energy* of a reaction. The activation energy of a reaction is the energy required to initiate a chemical reaction.

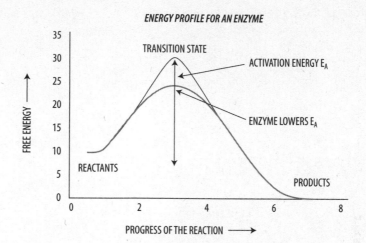

ENERGY PROFILE FOR AN ENZYME

i. When the enzyme and substrate are joined, a catalytic reaction takes place, forming a product. The enzyme can be recycled and used for later reactions.

2. The enzyme combines with the substrate or molecule that the enzyme will act upon. The shape of the enzyme's reactive site matches the shape of the substrate molecule.

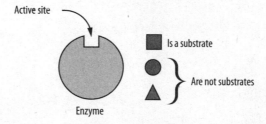

i. When the enzyme and substrate are joined, a catalytic reaction takes place, forming a product. The enzyme can be recycled and used for later reactions.
ii. Enzymes are affected by pH, temperature, and substrate concentration. Enzymes have a pH and temperature optima at which enzyme activity is greatest. Also, as a substrate's concentration increases, the speed at which the reaction occurs increases up to a maximum level at which all enzyme molecules are processing substrate molecules as fast as possible.
iii. *Cofactors* and *coenzymes* can also affect enzyme function. Sometimes the interaction between them causes a structural change, and therefore the enzyme's activity rate changes. Enzymes may also become active only when all necessary coenzymes and cofactors are present.

Cellular Structure

I. Key Concepts

A. All organisms are made up of one or more cells.

B. The cell is the basic unit of structure and function of organisms.

C. New cells arise only from existing cells by cell reproduction.

D. Cells exchange substances with their environment by transporting these substances in and out of the cell across the plasma membrane.

II. Types of Cells—Prokaryotic and Eukaryotic

A. *Prokaryotic cells* are simpler and more ancient than eukaryotic cells. The following are prokaryotes' important characteristics:

1. No nucleus, only a nucleoid region with one, circular DNA.
2. No membrane bound organelles.
3. Have a cell wall.
4. Like eukaryotes, prokaryotes contain a plasma membrane, cytoplasm, and ribosomes (location of protein synthesis).
5. No histones or no formation of chromosomes.

B. *Eukaryotic cells* are those of protists, fungi, plants, and animals. They contain the following characteristics:

1. Contain a nucleus, a nuclear envelope to protect DNA, nuclear pores to allow transport into and out of the nucleus, and linear DNA.

2. Have membrane-bound organelles. Plants have chloroplasts, for example, where photosynthesis is carried out, and many plant cells have a large, central vacuole that is absent in most animal cells.

3. Plants have rigid cells walls made of the polymer, cellulose, but animal cells do not have cell walls.

4. Like prokaryotes, eukaryotes contain a plasma membrane, cytoplasm, and ribosomes (location of protein synthesis).

5. Contain histones that form into chromosomes.

 III. **Eukaryotic Organelles**

A. Eukaryotic cells maintain internal organelles for specialized functions. Some of these include the following:

Feature	Structure	Function
Mitochondrion (-dria, pl.)	Small organelle with two membranes; inner membranes called *cristae* are folded to increase surface area for electron transport; directly requires oxygen	Site of aerobic respiration; produces ATP; inheritance is always from mother to child
Endoplasmic Reticulum (Rough and Smooth)	Rows of flattened, membranous sacs with or without ribosomes attached Rough ER-has ribosomes Smooth ER-no ribosomes	Sites of protein and membrane synthesis, including detoxification of drugs
Golgi Apparatus	Rows of flattened, membranous sacs	Modifies and transports proteins, etc., for export from the cell; synthesizes carbohydrates
Ribosome(s)	Tiny organelles; no membrane; contain rRNA and protein; bound to ER or float free in cytoplasm	Sites of protein synthesis

(continued)

Feature	Structure	Function
Lysosome(s)	Small, spherical; surrounded by one membrane; contains hydrolytic enzymes	Aids in phagocytosis and intracellular digestion
Vacuoles	Small or large; surrounded by single membrane	Provides turgor pressure for gross plant support; storage of substances
Chloroplast (Type of Plastid)	Various membrane bound organelles; chloroplast has double membrane plus thylakoids shaped like stacked coins to increase surface area	Site of photosynthesis; other plastids store starch or fats
Cytoskeleton	Network of microfilaments and microtubules throughout the cytoplasm	Controls cell shape; causes movement of chromosomes and organelles within the cell
Vesicle(s)	Small, spherical, numerous; surrounded by one membrane	Move substances from the ER to the Golgi apparatus and from there to the plasma membrane
Cilia and flagella	Hairlike; cilia are short and flagella are longer; 9 + 2 arrangement of microtubules	Locomotion of cells; movement of fluid surrounding a cell
Nucleus	Large, round; surrounded by nuclear envelope consisting of two membranes studded with pores	Site of chromosome (DNA) storage and RNA synthesis (transcription)
Nucleolus	Dense, spherical area within the nucleus	Site of rRNA synthesis and ribosome production
Cell Wall	Rigid; contains cellulose	Provides support and protection of cells

Membranes and Transport

I. Cell Membrane Structure

A. *Plasma Membrane*—separates internal environment from external environment and allows substances to be transported in and out of the cell.

B. *Selective Permeability*—the plasma membrane is selectively permeable, meaning that it allows some substances to pass through it, but not others. It is a direct consequence of the membrane structure, called the *Fluid Mosaic Model*. Essentially, the membrane is a mosaic of proteins that are embedded in or attached to the phospholipids.

1. The lipid portion of the membrane is composed mainly of phospholipids.
   i. *Phospholipids* have a *hydrophobic* (water-fearing) tail and a *hydrophilic* (water-loving) head.
   ii. The cytosol and the fluid outside the cell (the extracellular fluid) are both aqueous (watery) environments.
   iii. Therefore, phospholipids form a *bilayer*, as shown below, because the hydrophilic heads associate with the cytoplasm and the extracellular fluid, while the hydrophobic tails associate with each other.

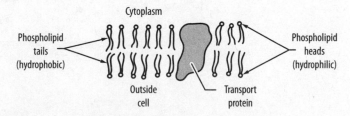

Cytoplasm

Phospholipid tails (hydrophobic)

Phospholipid heads (hydrophilic)

Outside cell

Transport protein

2. *Cholesterol* is also found in membranes and helps to keep the membrane fluid; they also lower the temperature required to make the membrane solid.

3. *Embedded proteins* in the cell membrane can be hydrophilic, with charged and polar side groups, or hydrophobic, with nonpolar side groups.

   i. The two types of proteins found in membranes are:
- ➤ *Integral Proteins*—transmembrane proteins with hydrophobic and hydrophilic portions
- ➤ *Peripheral Proteins*—bind to integral proteins on the outside of the cell membrane

   ii. Functions of membrane proteins include:
- ➤ transport
- ➤ enzymatic activity
- ➤ signal transduction and cell communication

II. Cell Membrane Transport

A. There are two main types of *cellular transport*: passive transport and active transport.

1. *Passive transport* does not require the cell to use ATP energy and plays a role in both the import of resources and the export of waste.

   i. In *diffusion*, a substance moves down its concentration gradient from an area of higher concentration to an area of lower concentration.
- ➤ Substances moved are small, uncharged molecules (e.g., carbon dioxide and oxygen).
- ➤ Substances move directly across the lipid bilayer.

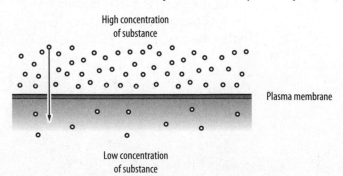

High concentration of substance

Plasma membrane

Low concentration of substance

ii. In *facilitated diffusion*, transport proteins move charged
molecules (e.g., potassium ions) and larger molecules
(e.g., glucose) into and out of the cell.
   ➤ As with diffusion, facilitated diffusion moves a
   substance down its concentration gradient from
   an area of higher concentration to an area of lower
   concentration without the use of ATP.
   ➤ Unlike diffusion, however, the substance moves with
   the help of carrier proteins or through a channel
   protein.

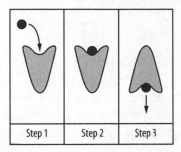

Carrier protein

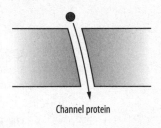

Channel protein

   ➤ Examples of facilitated diffusion include: glucose
   transport and Na^+/K^+ transport.
iii. *Osmosis* causes water to move across the plasma
membrane from a hypotonic solution to a *hypertonic
solution*.
   ➤ *Hypotonic solution*—has lower concentration of
   solutes (dissolved substances) than a hypertonic
   solution.
   ➤ If two solutions have equal concentrations of solutes,
   they are called isotonic and there is no net movement
   of water across the plasma membrane. This is called
   dynamic equilibrium.
   ➤ In osmosis, the solute molecule is not able to cross
   the *selectively permeable* plasma membrane.

Osmosis: Hypotonic, Hypertonic, and Isotonic Cells

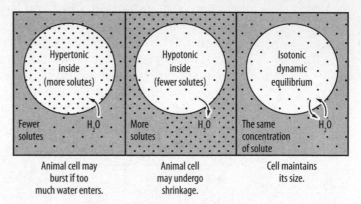

| Animal cell may burst if too much water enters. | Animal cell may undergo shrinkage. | Cell maintains its size. |

2. *Active transport* requires the cell to use ATP.
    i.   Involves movement against the concentration gradient, which is why active transport requires the use of ATP.
    ii. Specific membrane proteins are used in active transport.
    iii. Na$^+$-K$^+$ pump is an example of a protein that uses active transport to move ions through the cells' membrane.
3. *Exocytosis* and *endocytosis* move large molecules and food particles across the plasma membrane with the expenditure of ATP; in other words, they utilize active transport.
    i.   *Exocytosis*—fusion of vesicles and molecules with the plasma membrane; secretes materials to the outside of the cell.
    ii. *Endocytosis*—the cell takes in the molecules via vesicles that fuse with the plasma membrane.
    iii. *Pinocytosis*—uptake of liquids
    iv. *Phagocytosis*—uptake of solids

Homeostasis

I. Key Concepts

A. Each cell of a multicellular organism has the same genome, but cells differ from each other because they express different genes of the genome.

B. The body of a multicellular animal is organized into groups of cells called tissues, groups of tissues called organs, and groups of organs called organ systems.

C. Plants are multicellular, eukaryotic, autotrophic organisms organized into three main types of tissue systems—dermal, ground, and vascular. Each of the plant organs—roots, leaves, and stems—contain different structural arrangements of each tissue system.

D. Body systems coordinate their activities (e.g., heart rate and respiration rate increase if the muscles need more oxygen or a plant begins to preserve its use of water during times of water limitation).

E. *Homeostasis*—the maintenance of stable internal conditions in the body—is generally controlled by a group of body sensors, the nervous system, and the endocrine system that control in unison several body systems (e.g., thermoregulation involves the hypothalamus sensing body temperature, causing shivering of muscles and constriction of blood vessels to the skin, which produces sweat if the body is hot).

F. The structure of a component of the body or plant system underlies its function.

II. Overview of Coordinated Cooperation

A. The bodies of most animals are organized on four basic levels of coordinated cooperation: *cells, tissues, organs, and organ systems.* Plants are organized into three basic levels of coordinated cooperation: *cells, tissues, and organs.*

1. *Cells*—the most basic units of organization of all animals and plants.
2. *Tissues*—composed of many cells of the same type. For animals, they fall into four major categories:
   i. *Epithelial tissue*—functions include protection (e.g., skin), absorption and secretion (e.g., lining of the small intestine), and rapid diffusion (e.g., walls of capillaries and alveoli).
   ii. *Connective tissue*—create an extracellular matrix; examples include bone cells; cartilage, tendons, and ligaments; and blood and lymph cells.
   iii. *Muscle tissue*—capable of contraction and able to move substances; examples include skeletal muscles, smooth muscles, and cardiac muscles.
   iv. *Nervous tissue*—consists of nerve cells that receive and transmit electrical and chemical signals to provide coordinated communication between different parts of the body.
      ➤ For a list of plant tissues, see Section IV in this chapter, Plant Homeostasis through Specialization and Cooperation.
3. *Organs*—consist of an integrated group of different types of tissues that coordinate to perform specific functions in both animals and plants.
4. *Organ systems*—consist of a group of organs, and other related structures, that work together to carry out an overall process such as digestion or excretion in animals.
   i. Different organs not only interact with one another, but also coordinate specific biological functions. Some

examples of organ coordination are between the stomach and small intestine in order to digest food (animals); between the kidney and the bladder in order to excrete waste from the body (animals); and between the roots, stems, and leaves (plants).

ii. Organ system processes also work together in a coordinated fashion to allow integration of the body's activities.

III. Animal Homeostasis through Organ Specialization and Cooperation

As you review this section, pay particular attention to how the specialization of each organ is in coordination with other organs within that same system and also how organs from different systems work in concert to achieve homeostasis within the body. This topic is of particular interest on the AP Biology exam.

A. *Digestive System*—collection of structures, organs, and glands that work in concert to ingest food, break it into molecules that can be absorbed by the circulatory system, and eliminate solid waste from the body.

1. *Oral Cavity*—only carbohydrates are broken down
   i. *Mouth*—secretes *salivary amylase* which breaks down starch; chewing or mechanical digestion is also carried out; a *bolus* or ball of food is formed.
   ii. *Pharynx*—back of the throat; it has a structure called the epiglottis that blocks food from going down the windpipe or trachea.
   iii. *Esophagus*—food tube that transports bolus down to the stomach via a smooth muscle contraction called *peristalsis*.
2. *Stomach*—only protein is broken down
   i. *Gastric Juice*—a digestive fluid with pH of about 2 that aids digestion.

      ii. *Pepsin*—a protease secreted in an inactive form called pepsinogen until food is present in the stomach.

      iii. *Acid Chyme*—food and gastric juice that is processed in the stomach.

   3. *Small Intestine and Accessory Organs*—all 3 macro-molecules (carbohydrates, lipids, and proteins) are broken down:

      i. Organ that digests most food and absorbs it into the blood.

      ii. *Duodenum*—first part of the small intestine where digestion takes place.

      iii. *Pancreatic enzymes* are sent to the small intestine to aid in digestion. These enzymes are protease, amylase, and lipase. *Bile* from the liver (stored in the gallbladder) emulsifies or opens up the fat for lipase to break it down. Fat is broken down for the first time in the small intestine.

      iv. *Microvilli* of the small intestine increase the surface area and allow for absorption of nutrients.

   4. *Large Intestine or Colon* (no digestion)—main purpose is to reabsorb water; also creates feces and eliminates them through the rectum or end of the large intestine.

   5. Digestion involves both mechanical digestion (i.e., breaking down the food by physical action) and chemical digestion (i.e., breaking down food through the use of enzymes and other chemical means).

B. *Circulatory System*—Nutrients absorbed by digestion, wastes to be removed by the excretory system, oxygen gains through respiration, immune system components, hormones secreted from endocrine glands, and lymph fluid are the prime substances moved through the *circulatory system* from one part of the body to another.

   1. The *cardiovascular system* includes the heart, blood vessels, and blood.

   2. The *lymphatic system* includes lymph vessels, lymph nodes, and lymph fluid.

   3. There two types of circulatory systems:

      i. *Open Circulatory System*—blood mixes with internal organs directly. Insects, arthropods, and mollusks have an open circulatory system.

    ii. *Closed Circulatory System*—blood is contained within blood vessels that lead to the organs. Earthworms, octopi, and vertebrates have a closed circulatory system.

4. The structure of the circulatory system varies for different types of animals. For example:

    i. Fish—one ventricle, one atrium with gill capillaries allowing for gas exchange.

    ii. Amphibian—one ventricle and two atrium; have lung and skin capillaries for gas exchange. Have *double circulation* or oxygen-rich blood going to the organs and oxygen-deficient blood returning to the right atrium.

    iii. Mammal—two ventricles and two atrium, with lung capillaries for gas exchange. Also makes use of double circulation.

5. Major Components of the Human Circulatory System:

    i. *Heart*—Pumps blood

    ii. *Arteries*—Carry blood away from heart under pressure

    iii. *Veins*—Carry blood toward heart; contains valves that prevent blood from flowing backward; contraction of skeletal muscles provide force

    iv. *Capillaries*—Exchange substance with nearby tissues, generally by diffusion

    v. *Blood*

       ➤ *Red blood cells*—Carry O_2

       ➤ *White blood cells*—Protect against infections

       ➤ *Platelets*—Clot blood

       ➤ *Plasma*—Carries nutrients, wastes, antibodies, and hormones; regulates osmotic pressure

    vi. *Lymph Vessels*—Carry lymph from tissues to heart; transport digested fat to cardiovascular system

    vii. *Lymph Nodes*—Filter lymph to remove microorganisms and debris

    viii. *Lymph*—Lost fluid and protein is returned to the cardiovascular system

6. Flow of Blood in Mammalian Heart

    i. Deoxygenated blood from the vena cava enters the right atrium and passes through the right atrioventricular valve (tricuspid valve) into the right ventricle.

       ii. From the right ventricle it travels out of the heart via the pulmonary artery where it becomes oxygenated at the lungs (CO_2 is exchanged for O_2).

      iii. Oxygenated blood then travels via the pulmonary vein to the left atrium and through the left atrioventricular valve (mitral or bicuspid valve) into the left ventricle.

      iv. Oxygenated blood then leaves the left ventricle of the heart via the aorta to the organs of the body.

7. Beating of the Heart
    i. Cardiac muscle transfers an electrical signal via the following: Sinoatrial (SA) node or "pacemaker" of the heart (top of right atrium) generates an electrical impulse that is relayed to the atrioventricular (AV) node located between the right atrium and right ventricle.
    ii. The signal then transfers to the bundle branches and Purkinje fibers at the heart's apex.

8. Blood Pressure
    i. The force of blood against a blood vessel wall is a measure of blood pressure.
    ii. *Systolic pressure* is peak pressure in the arteries, which occurs when the ventricles are contracting.
    iii. *Diastolic pressure* is minimum pressure in the arteries, which occurs when the ventricles are filled with blood. The ratio of systolic and diastolic pressure is the measurement of blood pressure (systolic pressure/ diastolic pressure).

C. *Respiratory System*—consists of the lungs and related structures; it delivers oxygen to, and removes carbon dioxide from, the circulatory system in humans via the functional units known as the *alveoli*.

1. *Gas exchange*—defined as the uptake of oxygen (O_2) and loss of carbon dioxide (CO_2).
2. *Gills*—specialized for gas exchange of aquatic organisms.
3. *Tracheal system and lungs*—specialized for gas exchange of terrestrial organisms. For an insect such as a grasshopper, the tracheae opens to the outside.
4. *Lungs*—Amphibians (frogs) are the only vertebrates that use skin along with lungs to promote gas exchange.

5. Flow of Air in Mammalian Lungs
   i. Inhaled air passes through the *larynx* (upper part of the respiratory tract) into the *trachea or windpipe* (rings of cartilage).
   ii. The trachea divides into two *bronchi* leading to the lungs. The bronchi branch to *bronchioles* which contain air sacs called *alveoli*. The alveoli are covered with capillaries where the gas exchange takes place.
6. Control of Breathing
   i. *Medulla oblongata*, the lower part of the brainstem, maintains homeostasis by monitoring CO_2 levels.
   ii. When CO_2 levels are high, the CO_2 reacts with water in the blood, dropping the pH of the blood. The medulla oblongata senses a pH drop and excess CO_2 is exchanged for O_2.
7. Hemoglobin
   i. Iron—containing protein of mammalian red blood cells.

D. *Excretory System*—maintains water, salt, and pH balance, and removes nitrogenous wastes (urea in humans) from the body by filtering the circulating blood.

1. Major Components of the Excretory System—
   i. *Renal Artery*—carries unfiltered blood from the circulatory system to the kidneys for filtration.
   ii. *Renal Vein*—carries filtered blood from the kidneys back to circulatory system.
   iii. *Kidneys (contain nephrons)*—remove urea and toxins from the blood; produce urine; maintain salt, pH, and water balance of blood.
   iv. *Liver* (role in excretion)—synthesizes urea from ammonia; detoxifies other wastes.
   v. *Ureters*—carry urine from kidneys to bladder.
   vi. *Bladder*—stores and eliminates urine from the body.
   vii. *Urethra*—carries urine from the bladder to exterior of the body.
2. *Nephrons* in the kidneys filter blood, removing wastes and returning vital substances to the circulatory system. This is an example of coordinated cooperation.

3. Salt, water, and pH balance of the blood is maintained by the excretory system.

4. Urine is highly concentrated so that only a percentage of the water that enters the nephrons actually leaves the body along with wastes.

E. *Endocrine System*—uses hormones as long-distance cell-to-cell communication signals that allow different systems of the body to coordinate their activities to achieve a whole-body response to an event, or to maintain homeostasis.

1. Coordination between cells involving chemical messengers can be local or long distance. Examples include: neurotransmitters in the nervous system, histamine in the immune system, and prostaglandins in the immune and reproductive systems.

2. When a hormone comes in contact with a target cell, it either enters the cell or binds to receptors on the surface of the cell.

3. The human body has exocrine and endocrine glands.
   i. *Exocrine glands*—such as the salivary glands, deliver their secretions through ducts.
   ii. *Endocrine glands*—secrete their hormones directly into the interstitial fluid surrounding the gland, where they are picked up readily by the bloodstream.
   iii. Some glands, such as the *pancreas,* have both endocrine and exocrine functions (e.g., the exocrine function of the pancreas is digestion, as well as the release of the endocrine hormone, insulin).

4. Some of the Major Components of the Endocrine System—
   i. *Hypothalamus*—regulates pituitary gland
   ii. *Posterior Pituitary*—uterine contractions and lactation; water reabsorption in the kidneys
   iii. *Thyroid*—stimulates metabolism by multiple effects; decreases calcium levels in the blood
   iv. *Thymus*—T-cell development (immune system)
   v. *Ovaries*—growth of uterine lining; initiate and maintain secondary sex characteristics
   vi. *Testis*—sperm formation; initiate and maintain secondary sex characteristics

vii. *Lining of small intestine/stomach*—stimulates small intestine/stomach to release enzymes

5. The interplay between the nervous system and the glands of the endocrine system coordinate body functions in other organ systems.

6. The same hormone can have multiple effects if it targets more than one type of cell, tissue, or gland.

7. *Homeostatic regulation* often involves two negative feedback loops and two antagonist hormones.

8. Positive feedback in the endocrine system involves amplifying the production of a hormone until the hormone reaches sufficient levels to precipitate an event.

F. *Nervous System*—receives input from internal and external sensors and relays that information to the brain, where integration occurs; the brain sends out nervous signals to different parts of the body that carry out actions in response.

1. The functional units of the nervous system are nerve cells (*neurons*). Neuron structures are as follows:
   i. *Cell body*—large portion of neuron that contains the nucleus and organelles.
   ii. *Dendrites*—communicate the nervous signals from tips of neuron to the cell body.
   iii. *Axon*—transmit action potentials down their lengths.
   iv. *Myelin Sheath*—insulate axons for faster action potential in the central nervous system.
   v. *Schwann Cells*—insulate axons for faster action potential in the peripheral nervous system.
   vi. *Nodes of Ranvier*—gaps between the Schwann cells that allow for faster action potentials, hence faster communication.
   vii. *Synapse*—the space between the end of the axon and the target; examples of targets include muscles or other neurons.
   viii. *Axon terminals*—convert action potential into chemical signal when an action potential triggers the release of its vesicles filled with neurotransmitters into the synaptic cleft.

        ix. *Synaptic Cleft*—Provides location for transmission of nerve signal between two neurons.

2. The nervous system is divided into a *central nervous system* and *the peripheral nervous system*:
   i. *Central nervous system*—consists of the brain and spinal cord.
   ii. *Peripheral nervous system*—consists of all the neurons outside the brain and spinal cord, and it has two main divisions:
      ➤ *Sensory division*—brings information from sense to organs to the central nervous system via afferent (incoming) neurons.
      ➤ *Motor division*—brings information from the brain to the body by efferent (outgoing) neurons, and is divided into voluntary and involuntary systems.

3. Special types of neurons, called *sensory receptors*, detect stimuli and turn it into action potentials that travel to the proper region of the brain where the received sensation is processed to produce a perception (such as a visual image we might recognize as an apple or a banana).

4. Receptor types include *photoreceptors, mechanoreceptors, chemoreceptors, thermoreceptors,* and *pain receptors*.

5. The Na^+-K^+ pump maintains a negative charge inside a neuron.

6. When a neuron is stimulated, this charge difference is reversed by the opening of ion channels in the plasma membrane, causing a rush of positive Na^+ into the cell and is the source of action potential, which is a wave of depolarization that travels along the axon's plasma membrane.

7. *Sense organs* detect stimuli, convert it to action potentials, and send it to the brain for interpretation:
   i. *Eyes*—sense differences in light intensity and wavelength using photoreceptors.
   ii. *Ears*—sense differences in pressure using mechanoreceptors for hearing and balance.
   iii. *Nasal passage and mouth*—contain chemoreceptors that detect different types of chemicals.
   iv. Mechanoreceptors, heat and cold receptors, and pain receptors detect touch, temperature, and tissue damage.

G. *Skeletal, Muscular, and Nervous Systems*—coordinate the movement of body parts and the process of locomotion.

1. The skeletal system of humans is an *endoskeleton* composed of bones joined together at joints. The skeleton is composed of two parts: the *axial skeleton* (contains the skull, ribcage and spine) and the *appendicular skeleton* (contains all the other bones).

2. *Bones* are living tissues containing mineral deposits in specific arrangements interspersed with bone cells—osteocytes.

3. Bones of the skeleton have a variety of functions:
   i. Storage sites for minerals such as calcium and phosphorous.
   ii. Provide protection and support for softer internal structures (e.g., skull and ribcage)
   iii. *Red bone marrow* in parts of the sternum, pelvis, ribs, and ends of the long bones produce red blood cells and white blood cells.
   iv. *Yellow bone marrow*, inside the shafts of long bones, stores energy in fat cells.
   v. Provide attachment sites for muscles and provide solid support against which muscle contraction acts to cause movement of different parts of the body.

4. *Joints*—regions where bones are attached to each other by ligaments.
   i. Some types of joints include: fixed, semimovable, and movable.

5. *Skeletal muscles*—attached to bones by tendons—cause voluntary body movements and locomotion in humans in response to nerve impulses when their functional units, called sarcomeres, contract and relax.
   i. Pairs of opposing muscles—flexors and extensors—control the movement of limbs.
   ii. Contraction of millions of sarcomeres within a muscle causes the whole muscle to contract.

H. *Immune System*—specific and nonspecific defenses used by the body to fight pathogens.

1. *Pathogens*—organisms that cause infectious disease, including viruses, bacteria, protists, fungi, and small vertebrates.

2. *Nonspecific defenses*—barriers or systems that attack all pathogens regardless of their type. These include:
   i. *Skin*—provides physical and chemical protection.
   ii. *Mucus*—traps pathogens that enter epithelia of systems that open to the exterior (e.g., trachea, bronchi, urethra, vagina)
   iii. *Cilia*—sweep pathogens to exterior opening of respiratory tract.
   iv. *Stomach acid*—swallowed pathogens are destroyed by the low pH of the stomach.
   v. *Inflammatory response*—histamine is released and causes increased blood flow to damaged area, allowing specific immune cells to destroy the pathogens.
   vi. *Interferon*—inhibits reproduction of viruses.
   vii. *Fever*—suppresses bacterial growth and stimulates the immune system.

3. Organs of the immune system include: *bone marrow,* the *thymus, lymph nodes, tonsils, adenoids,* and the *spleen.*

4. *Specific defenses*—those targeting specific pathogens by recognizing a specific foreign substance by its antigens and then marshaling the humoral and/or cell-mediated defenses.
   i. *Humoral immune response*—involves B cells, which attack pathogens with antibodies.
   ii. *Cell-mediated response*—involves T cells, which attack pathogens, cells infected with pathogens, and cancer cells by lysing them.

IV. Plant Homeostasis through Specialization and Cooperation

As you review this section, pay particular attention to how the specialization of each cell, tissue, and organ is in coordination with each other in order to achieve homeostasis. This topic is of particular interest on the exam.

A. Plants are organized into cells, tissues, and organs that are specialized in order to carry out necessary functions for the plant's survival. Cooperation of these elements is essential to survival success.

B. *Cells*—three types of cells are found throughout the plant in different tissues and have different functions consistent with their structures:

1. *Parenchyma*—used in storage, photosynthesis, and protection and transport.
2. *Collenchyma*—supports growing parts of plant/growing regions of the stem.
3. *Sclerenchyma*—supports non-growing parts of the plant; transport/fibers, and vessels of vascular tissue.

C. *Tissues*—three types of tissue systems are present in all organs throughout the plant:

1. *Dermal tissue system*—is the outermost layer of the plant and serves as protection as well as allowing gas exchange and mineral and water absorption.
2. *Ground tissue system*—comprises the bulk of plant roots and offers support, storage, and photosynthesis/fibers.
3. *Vascular tissue system*—contains cells specialized for transport that are bundled into groups and are embedded in ground tissue; transports water and products of photosynthesis. This system contains several important cell types:
   i. *Xylem*—transports water from the roots through the stem and to the leaves; involved in the process called transpiration.
   ii. *Phloem*—transports nutrients and hormones from sources to sinks within living systems.

D. Another important tissue in plants is *meristematic tissue*, which contains cells capable of dividing to produce new tissues throughout the life of the plant.

E. Organs—consist of an integrated group of different types of tissues that coordinate to perform specific functions. Plants possess three different organs:

1. *Leaves*—primary function is to provide nutrition and carry out photosynthesis.
2. *Roots*—primary function is to anchor the plant, to absorb water and mineral nutrients, and to store carbohydrates.
3. *Stems*—primary function is to support the shoot and to transport water and sugar.

F. Hormones—plants sense and respond to changing conditions in the environment and exert internal regulation over growth and development through the action of hormones (sometimes called growth regulators in plants).

1. Plant hormones are similar to animal hormones in important ways, and a single plant hormone can have different effects depending on a number of factors.
2. The five major types of plant hormones are: *auxin*, *cytokinins*, *ethylene*, *abscisic acid*, and *gibberellins*.
3. Plant movements are one way a plant responds to its environment and can be the slow results of changing growth patterns or rapid movements that involve reversible changes in specific cells.
4. Plants respond to seasonal changes by sensing photoperiod, temperature, or a combination of the two.

G. Immune Response—plants can also sense the presence of pathogens, mainly through the damage they cause to plant cells; and plants respond by producing compounds called *phytoalexins* and *pathogenesis-related proteins*, that help protect the plant.

V. Negative and Positive Feedback

A. Organisms use both negative and positive feedback mechanisms in order to maintain their internal environments, respond to external stimuli, and regulate growth and reproduction.

1. *Negative feedback*—occurs when a stimulus produces a result, and the result inhibits further stimulation

> Examples include temperature regulation in animals and a plant's responses to water limitations.
2. *Positive feedback*—occurs when a stimulus produces a result, and the result causes further stimulation, thereby triggering an event.
   > Examples include lactation in mammals, the onset of childbirth, and the ripening of fruit.

B. Alterations in this feedback can cause detrimental consequences for the organism.
   > Examples include diabetes in response to decreased insulin, Graves' disease (hyperthyroidism), blood clotting, and dehydration in response to antidiuretic hormone (ADH).

VI. Evolutionary Similarities and Examples

A. Homeostatic mechanisms are reflected in organisms of common ancestry. These mechanisms can continue to be similar in different organisms of different species over time, or they can change in order to help organisms maintain homeostasis in their specific environments.

B. Homeostatic mechanisms and continuity

1. *Excretory System* is found *in flatworms, earthworms,* and *vertebrates.*
   i. *Flatworms*—dispose of waste through tiny holes that are attached to internal tubes called *protonephridia.*
   ii. *Earthworms*—dispose of waste through tiny holes on their undersides that are attached to internal tubes called *metanephridia.*
   iii. *Vertebrates*—dispose of waste through two holes (anus and urethra) that are connected to tubes called *nephrons* within *kidneys.*
2. All three of these organisms demonstrate continuity in how they maintain homeostasis by disposing of waste from their bodies. The continuity in this specific homeostatic mechanism also demonstrates their common ancestry.

C. Homeostatic Mechanisms That Change Over Time

1. *Respiratory system* is found in both aquatic and terrestrial animals in the forms of *gills* (aquatic) and *lungs* (terrestrial).
    i. For the respiratory system to function, it requires a wet environment, large surface area, thin membranes, and the presence of O_2.
    ii. A*quatic environment*—gills are specifically customized to accommodate the organisms by acquiring these requirements of the respiratory system. Aquatic environments are extremely wet and have little O_2. Therefore, gills are on the outside of fish, for example, because they are able to be kept wet at all times by the surrounding water; they have a large surface area and thin membranes in order to effectively absorb as much O_2 as possible.
    iii. *Terrestrial environment*—lungs are specifically customized to accommodate the organisms by acquiring these requirements of the respiratory system. Terrestrial environments have little water but lots of O_2. Therefore, organisms with lungs have this organ tucked safely inside their bodies in order to preserve their wetness; lungs also have large surface areas and thin membranes in order to easily absorb O_2.
    iv. Although the same characteristics of the respiratory system exist for both aquatic and terrestrial organisms, over time, these organisms developed organs to help them "breathe" in their respective environments.

 VII. **Disruptions of Homeostasis**

A. Biological systems are affected by disruptions in their dynamic homeostasis. These disruptions can occur at the molecular level and affect the health of the organism or at the level of the entire ecosystem and affect the survival of parts or all of the population.

B. Molecular and Cellular Disruptions of Homeostasis

1. Foreign, toxic substances enter the organism and are detected by multiple cells, tissues, and organs in order to expel it from the organism and to reinstate homeostasis.
2. The body combats these disruptions through physiological and immunological responses.

C. Disruptions to an Ecosystem's Homeostasis—these disruptions can come from the following sources:

1. *Other species*—other species can disrupt the homeostasis of another species through predation or parasitism.
2. *Human impact*—humans can impact not only their ecosystems but also those of other species; an example includes contamination of a local lake that is not only a water source for humans but also home to fish, insects, and birds that are also affected by the contamination.
3. *Weather*—such as hurricanes, earthquakes, or floods
4. *Limitation of water*

Reproduction, Growth, and Development

I. Reproductive Process of All Land Plants

A. *Alternation of Generations*—sexual life cycle of all plants, although sexual reproduction specifically varies from group to group.

1. Unlike animals, in which the diploid organism is the only multicellular form, in plants there are two multicellular forms: the *gametophyte* and the *sporophyte*.
2. *Gametophyte*—haploid and produces the egg and sperm.
3. *Sporophyte*—diploid and is formed by the fusion of egg and sperm.
4. *Meiosis*—produces the spore that will eventually give rise to sporophytes via mitosis.
5. *Sporophyte*—diploid multicellular generation and produces haploid spores by meiosis; product of fertilization of male and female gametophytes.

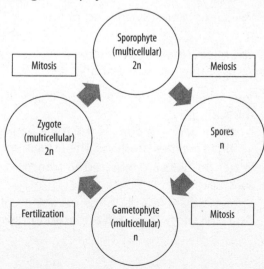

II. The Rise of Land Plants & Plant Reproduction

A. *Green algae*—gave rise to land plants; they mostly live in fresh water, but there are some marine ones as well. Green algae are not considered officially land plants. They have no vascular tissue or root system.

B. *Bryophytes*—include moss, liverworts, and hornworts. They are nonvascular and seedless, have no root system, but they are anchored via *rhizoids* or nonvascular containing cells.

1. The dominant stage during the lifecycle of bryophytes is the gametophyte.
   i.   *Antheridia*—produces the male gametophyte.
   ii. *Archegonia*—produces the female gametophyte.
   iii. Fertilization takes place in the archegonia. Water droplets are required to transport male gametophyte to archegonia.
   iv. *Sporangium* produces spores that will eventually produce the mature male or female gametophyte.

C. *Tracheophytes*—have four major characteristics:

1. Protective layer that surrounds the gametes.
2. Multicellular embryos.
3. Cuticles or waxy layer that covers all parts above the root system.
4. Vascular system (xylem and phloem).

D. *Pteridophytes* (most basic tracheophytes)—include ferns and horse tails; they are vascular and seedless.

1. The dominant stage during the life cycle of pteridophytes is the sporophyte.
   i.   *Antheridia*—produces the male gametophyte.
   ii. *Archegonia*—produces the female gametophyte.
2. Water droplets are required to transport male gametophyte to archegonia.
3. Sporangia are found on the underside of the leaf and produce spores that will undergo fertilization.

E. *Seeds*—because some plants, unlike animals, are incapable of locomotion, seed and fruit dispersal are the main ways plants can migrate from one location to another.

　1. Some seeds, such as the winged seeds of maple trees and the fluffy seeds of dandelions and milkweed, are dispersed by *wind*.
　2. Other seeds are surrounded by *burrs* that are dispersed by attaching to the fur of mammals.
　3. Some plants have seeds surrounded by *fruit* that, when eaten and eliminated by animals, can widely disperse their seeds.

F. A seed contains an embryo and a food supply chain surrounded by a tough outer covering called a seed coat.

　1. *Monocots*—have one cotyledon and are compressed
　2. *Dicots*—have two cotyledons
　3. The embryo of a seed has four main parts:
　　i. *Radicle*—the embryonic root
　　ii. *Hypocotyl*—the stem from the radicle to the cotyledons
　　iii. *Epicotyl*—the stem above the cotyledons
　　iv. *Plumule*—a small group of embryonic leaves

G. *Gymnosperm*—conifers or cone-bearing plants such as pine.

　1. The dominant stage during the life cycle of gymnosperms is the sporophyte.
　2. *Ovule*—structure containing the eggs that are produced via meiosis.
　3. *Pollen Cone*—pollen grains (haploid) are produced via meiosis.
　4. *Ovulate Cone*—contains two ovules.
　5. *Pollen Grain*—2 male gametophytes formed via pollination land on ovulate cone. One will be destroyed while the other will wait at least one year before fertilization takes place.
　6. Once a new embryo (only 1 of the eggs is fertilized) is produced (sporophyte), a seed coat will be produced and the unfertilized female gametophyte will become the food reserve.

H. *Angiosperm*—flowering plant.

1. The dominant stage during the life cycle of angiosperms is the sporophyte.
2. Reproductive structure of angiosperms is the flower.
   i. *Sepal*—enclose and protect the flower before it buds.
   ii. *Petal*—bright-colored structure that attracts insects for pollination.
   iii. *Stamen*—male reproductive organ.
   iv. *Carpel (pistil)*—female reproductive organ.
   v. *Anther*—part of the stamen where pollen is produced via meiosis.
   vi. *Stigma*—part of the carpel; receives the pollen.
   vii. *Style*—part of the carpel that leads to the ovary.
   viii. *Ovary*—part of the carpel where the ovule is encased.
3. The programmed cell death (*apoptosis*) of the flower plays a normal role in the development of the plant; flower cell death also indicates that pollination/fertilization has already occurred.
   i. Unlike leaf loss, flower death is genetically programmed and not necessarily environmentally cued.
   ii. Pollination is an important trigger for flower cell death in many plant species.

I. Fruits—ripened ovaries.

1. After fertilization, the ovary thickens to aid in protection.
2. Fruits aid in dispersing seeds because some are edible by other species. Once eaten, the seed's protective coating will not break down in the organism, but the seed will pass through the organism's feces at a distant location.

J. Requirements for *Seed Germination*

1. All seeds need sufficient water before they will germinate, ensuring that there will be enough water for the developing seedling to survive.
2. The seeds of many plants adapted to temperate climates do not germinate until they have experienced a cold period of a specific duration.

3. Some seeds require *light* to germinate, ensuring that seeds buried too far underground will not germinate if they are too far away from the surface to survive.

4. Seeds also require *oxygen* for cellular respiration: when the seed coating has opened, and if the seed is close enough to the surface, it can obtain enough oxygen to germinate.

K. *Double Fertilization* and the *Endosperm*

1. In contrast to gymnosperms where one of the pollen grains is destroyed, both pollen grains fertilize one egg in angiosperms. This is called double fertilization.

2. The triploid nucleus will continually divide, giving rise to a rich food reserve called the endosperm.

3. *Cotyledon* or seed leaves are produced. Monocots produce 1 seed leaf, while dicots produce 2 seed leaves.

Be sure to know the four main plant groups: bryophytes, seedless vascular plants, gymnosperms, and angiosperms. The gametophyte is the dominant generation of bryophytes, while the sporophyte is the dominant generation in seedless vascular plants. In seed plants, such as gymnosperms and angiosperms, the seed replaces the spore as the main means of dispersing offspring.

III. Plant Growth and Development

A. Plant growth can occur in the following ways:

1. *Apical meristems or primary growth*—located at the tips of roots or shoot buds and contain the cells undergoing mitosis for vertical or expansive cell growth.

2. *Lateral meristems or secondary growth*—located through the length of the shoot system and roots and is considered outward horizontal growth (increases plant's diameter).

B. Cell types that assist in development and growth

1. *Parenchyma cells*—perform the metabolic processes of cells.

2. *Collenchyma cells*—support the plant.
3. *Sclerenchyma cells*—rigid cells that are found in areas where the plant is no longer growing.

C. Tissue types that assist in development and growth

  1. *Vascular-xylem tissue*
     i. Type of vascular tissue that transports water and dissolved minerals from the roots up the plant.
     ii. *Tracheids* and *vessel elements*—dead cells that conduct water and minerals.
  2. *Vascular-phloem tissue*
     i. Type of vascular tissue that transports food from the leaves to the roots of the plant.
     ii. *Sieve-tube members*—live cells, but have no organelles; their main function is to transport sucrose.
     iii. *Companion cells*—next to the sieve-tube member and provide all the metabolic resources for the sieve tube members.
  3. *Dermal tissue*—protects the plant.
     i. *Cuticle*—waxy coat that helps the plant retain water.
  4. *Ground tissue*—occupies space between vascular and dermal tissues; mostly made up of parenchyma cells.

D. *Adverse effects on growth and development*

  1. Environmental factors
     i. Various environmental factors, such as drought, extreme temperatures, or lack of sufficient nutrients available in the soil can all negatively affect the growth and development of plants.
     ii. If these adverse conditions are brief in time and not too extreme, then chances are that the plant will survive, although there might be leaf loss or a diminishment to the overall health of the plant.
     iii. Responses to adverse environment conditions:
        ➤ *Drought*—plants often close the stomata in order to reduce transpiration (loss of water through the plant's leaves)

➤ The purpose of transpiration is to cool the plant by pulling water and nutrients from the soil to parts of the plant, especially the leaves.

➤ Extended drought conditions cause the plant's water supply to diminish, which inhibits the plant from regulating its temperature; ultimately, the plant can become nutrient-deficient and photosynthesis can be compromised.

➤ *Extreme temperatures*—cause the plant to conserve its water supply and to increase transpiration in order to help stabilize the plant's temperature; if more water is unavailable, then extreme temperatures can negatively affect the health of the plant.

➤ *Leaf loss* is one way the plant can minimize the areas in which water must be dispersed in order to maintain the plant's overall temperature; in extreme heat or drought, leaf loss is one way the plant is able to "cope" and attempt to preserve itself.

➤ *Genetic mutations*—a genetic mutation is not always a bad thing for a plant; however, any mutation that compromises the plant's ability to obtain nutrients, to grow and develop, and to maintain homeostasis can have an adverse effect on the plant and possibly result in death.

IV. Animal Reproduction

A. Various types of reproductive patterns

1. *Asexual reproduction*—no genetic diversity since all genes come from one parent. No fusion of egg and sperm.
   i. *Budding*—outgrowths from a parent form and pinch off to live independently.
   ii. *Binary fission*—a type of cell division by which prokaryotes reproduce; each daughter cell receives a single parental chromosome.
   iii. *Fragmentation*—breaking of a body piece that will form an adult via regeneration of body parts.

2. *Sexual reproduction*—genetic diversity since genes will be inherited from both parents. Fusion of egg and sperm.
3. *Parthenogenesis*—egg develops without being fertilized.
4. *Hermaphroditism*—having both male and female reproductive organs.

B. *Spermatogenesis*—production of sperm; is stimulated by the hormone testosterone, and occurs in the seminiferous tubules of the testes, which are located in the scrotum.

1. Continuous throughout the life of a male.
2. Four viable sperm are produced during each meiotic division; meiosis occurs in an uninterrupted sequence, unlike in females.

C. *Oogenesis*—production of ova and happens in ovaries.

1. Egg cells begin with meiosis, but are arrested in *prophase* until one egg per menstrual cycle is stimulated to complete meiosis by the hormone, *FSH*; meiosis then stops again and the ova does not undergo *meiosis II* until after fertilization.
2. At a female's birth, the ovary contains all the cells that will develop into eggs from puberty to menopause.
3. Only one viable ovum is produced at a time with three non-viable polar bodies.

D. *Reproductive cycle of the human female*—also called the menstrual cycle or changes in the uterus or female reproductive organ; it's a 28-day cycle in which the destruction and regeneration of the uterine lining (endometrium) occurs.

1. *Menstrual phase* (day 0–5)—menstruation (bleeding due to destruction of endometrium).
2. *Proliferative phase* (day 6–14)—regeneration of endometrium.
3. *Secretory phase* (day 15–28)—endometrium becomes more vascularized and is ready for implantation of embryo. If the embryo is not implanted, the entire menstrual cycle will happen again.
4. *Ovarian cycle*—parallels the menstrual cycle.

    i. *Follicular phase* (day 0–13)—egg cell enlarges in a follicle.

    ii. *Ovulation* (day 14)—oocyte is released and pregnancy can take place.

    iii. *Luteal phase* (days 15–30)—the corpus luteum is formed, which is a structure that grows on the surface of the ovary where a mature egg was released at ovulation. The corpus luteum produces progesterone in preparing the body for pregnancy.

5. *Important hormones*—under the control of four endocrine hormones—*LH, FSH, estrogen, and progesterone*—one of two ovaries releases one egg cell per menstrual cycle.

    i. LH (luteinizing hormone) and FSH (follicle stimulating hormone) are made in the anterior pituitary.

    ii. Estrogen and progesterone are made in the ovaries.

    iii. *FSH*—stimulates ovulation to occur

    iv. *Estrogen*—causes cell division in the uterine lining and stimulates release of LH during menstruation; during the luteal phase, estrogen causes LH and FSH levels to fall.

    v. *LH*—when LH levels spike during menstruation, the follicle bursts, releasing the egg.

    vi. *Progesterone*—causes blood vessel growth in the uterine lining during the luteal phase.

V. Animal Development

A. *Fertilization*—process of the egg and the sperm fusing to make a zygote.

1. Activation of the egg will lead to embryonic development.

2. Fertilization usually occurs in the fallopian tubes of the female reproductive tract, where the head and midpiece of one sperm usually combines with one egg.

B. *After Fertilization*

1. After implantation of the fertilized egg, pregnancy ensues and a placenta is formed to provide nutrition from the mother to the developing fetus through the umbilical cord without the mixing of maternal and fetal blood.

2. Rapid *organogenesis* occurs in the fetus during the first trimester of pregnancy, making it especially important to protect the fetus from damaging substances that may harm it.

3. *Labor* produces strong contractions of the smooth muscles of the uterus, the cervix dilates, and the baby is pushed out through the vagina.

C. *Summary of core concepts in animal development*

1. Cell types in a multicellular organism are different from one another because they express different genes present in their identical genomes.

2. Developmental genes, such as the homeotic genes, generally code for proteins that regulate differential gene expression.

3. Development in animals involves cell division, cell differentiation, and movement of cells in the developing embryo.

4. Embryonic development in most vertebrates starts with a zygote that develops into a blastula and then a gastrula.

5. Most animals have three embryonic tissue layers: ectoderm, mesoderm, and endoderm.

6. Cells from animals with indeterminate cleavage can be separated from each other within the first few mitotic divisions, and each cell can then generally go on to form an entire organism.

7. The digestive tract of most animals develops from the archenteron and has two openings.

8. The coelom is the second internal compartment in many animals that cushions and protects internal organs.

PART IV

GENETICS AND INFORMATION TRANSFER

DNA Structure and Replication

Key Concepts

A. *Deoxyribonucleic Acid (DNA)*—genetic material

B. The structure of a DNA molecule is the key to understanding how each strand of DNA can act as a template for the replication of the other strand during DNA replication and for the production of RNA during transcription.

C. DNA replication is a *semi-conservative process* that produces two new DNA molecules, each of which consists of one old strand and one newly synthesized complementary strand, and which are checked for errors by proofreading and repair processes.

Discovery of DNA as the Genetic Material

A. Transformation Experiments of Griffith, Avery, McCarty, MacLeod.

1. Smooth (contains capsule) living *Streptococcus pneumonia* injected into live mouse; it resulted in a dead mouse.
2. Rough (no capsule) living *Streptococcus pneumonia* injected into live mouse; it resulted in a healthy mouse.
3. Heat-killed smooth (capsule destroyed) *Streptococcus pneumonia* injected into live mouse; it resulted in a healthy mouse.
4. Heat-killed smooth (contains capsule) mixed with living rough (no capsule) *Streptococcus pneumonia* injected into live mouse; it resulted in a dead mouse.

5. *Interpretation of the experiment*—DNA from the heat-killed smooth cells "transformed" the rough cells into smooth cells that killed the mouse. The transforming agent was DNA.

B. Hershey-Chase Experiment

1. Worked with T2 bacteriophage or a virus that infects bacteria.
2. Bacteriophage were radioactively labeled with P32 (DNA) or S35 (protein coat of bacteriophage).
3. When separate experiments were completed, it was found that bacteria contain the radioactively labeled P32 DNA of the bacteriophage.
4. *Interpretation of the experiment*—bacteriophage injected their DNA into the host bacterium in order to produce progeny phage, indicating DNA as the genetic material.

C. Watson and Crick

1. *James Watson, Francis Crick, Rosalind Franklin*, and *Maurice Wilkins* contributed to constructing the double helical model of DNA.
2. *X-ray crystallography*—a technique used to measure the shapes of molecules—was contributed by Franklin and Wilkins to determine that DNA was a double helix.

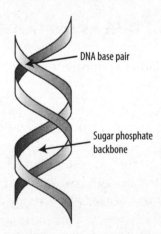

DNA base pair

Sugar phosphate backbone

3. Watson and Crick used this and other data to construct a *three-dimensional model.*
   i. Two strands of complementary DNA twist to form a *helix* often described as a spiral ladder with each base pair representing a single step.
   ii. Each turn of the double helix contains 10 base pairs and is 34 angstroms long.
   iii. The width of the helix is uniform and is 10 angstroms across.
4. Base pairing rules of purines and pyrimidines were established (also known as Chargaff's rule).

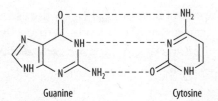

Hydrogen bonds indicated by dashed lines.

Adenine   Thymine

Guanine   Cytosine

i. Adenine (purine) pairs with thymine (pyrimidine). 2 hydrogen bonds for base pairing.

ii. Guanine (purine) pairs with cytosine (pyrimidine). 3 hydrogen bonds for base pairing.

D. Meselson-Stahl

1. Experiment indicated that replication of DNA is semi-conservative, or one old strand is used for the synthesis (template) of a new strand.
2. Experiment showed that both heavy and light nitrogen would be incorporated into the daughter DNA during the first round of DNA replication. In the second round of replication, daughter strands would have only light nitrogen since the heavy nitrogen was removed. Banding patterns indicated a semi-conservative model is favored over conservative or dispersive.

Test Tip

Every AP Biology test-taker should know the main differences between DNA and RNA. DNA consists of A, T, C, G as nitrogenous bases, deoxyribose as 5 carbon sugar, and phosphate. RNA consists of A, U, G, C as nitrogenous bases, ribose as 5 carbon sugar, and phosphate. Structurally DNA is a double-stranded helix, while RNA is single-stranded.

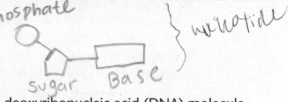

Phosphate

nucleotide

sugar   Base

III. DNA Structure

A. The structure of a deoxyribonucleic acid (DNA) molecule is based on the pairing of nucleotides along the lengths of two complimentary DNA strands, each of which has a sugar-phosphate backbone and twists to form a double helix.

• fits more info into a smaller space
• protects vital info

DNA Quiz: 2/3/15

B. *Sugar-phosphate backbone*—of each DNA strand is a repeating chain of the 5-carbon sugar, *deoxyribose*, and a *phosphate group* composed of a phosphorus atom surrounded by 4 oxygen atoms.

1. The sugar-phosphate chains are held together by *covalent bonds* that are generally only broken or formed by *enzymes*.
2. The sugar-phosphate backbone of each strand is an identical feature of all DNA molecules.

C. *Nitrogenous bases*—attached to each sugar is one of the four nitrogenous bases composed of carbon and nitrogen rings.

1. *Purines*—adenine (A) and guanine (G), consist of two rings.
2. *Pyrimidines*—thymine (T) and cytosine (C) have a single ring.

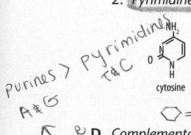

Purines > Pyrimidines
T & C
A & G
↑
both have "y"

NH₂

NH₂

HN, CH₃

HN

cytosine   adenine   thymine   guanine

⬡ = pyrimidines   ⬡⬡ = purines

D. *Complementary pairing*—the complementary pairing of nitrogenous bases is the basis of the double-stranded structure of a DNA molecule.

A & T → pointy
C & G → curvy

1. *Base pairing* involves the formation of *hydrogen bonds* between the bases that are located toward the center of each DNA molecule, and they serve to hold the two strands together.

   easier separation

   i. Hydrogen bonds are weaker than covalent bonds, allowing the two strands of the DNA molecule to be separated for DNA replication and transcription.

   ii. A pyrimidine can only pair with a purine due to their sizes and the types of hydrogen bonds possible between the bases. *1 pyrimidine & 1 purine = 1 base pair*

2. Base pairing occurs along the entire length of the two DNA strands and results in one strand being an exact complement of the other.

3. Knowing the sequence of one strand of DNA, the sequence of its complement can be deduced by the use of the base pairing rules.

   -----G A T T C G T A A G G C------ one strand of DNA
   -----C T A A G C A T T C C G------ complementary strand of DNA

   ✱ Know the opposites

Test Tip

An understanding of the similarities and differences between DNA replication, transcription, and translation is a popular topic for test questions.

Chargaff's Rule
A 1:1 ratio of A's & T's
A 1:1 ratio of G's & C's

Ex) A=40%   T=40%
C=10%   G=10%

IV. DNA Replication

A. During replication, each strand of a DNA molecule acts as a template for the synthesis of the other strand, and when errors do occur, proofreading and repair mechanisms keep mutation rates low.

B. All of the chromosomes of an organism's *genome* (all of an organism's DNA) are copied prior to each cell division.

   1. Each strand acts as a *template* for the synthesis of its complementary strand through the addition of nucleotides to the growing end of the complement.

2. DNA replication begins at *origins of replication* distributed at many sites along the length of each eukaryotic chromosome and usually at a single site on the chromosome of prokaryotes.

3. *DNA polymerase* is the enzyme that catalyzes the addition of nucleotides to the ends of a growing strand of DNA through a process called elongation.

C. The process of DNA replication occurs as follows:

1. Enzymes, called *helicases*, unwind the helix at the origins of replication and help break the hydrogen bonds holding the strands together, creating a *replication fork*.

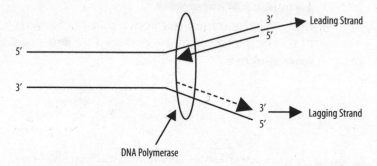

DNA Polymerase

i. *Leading Strand*—the daughter strand that is synthesized into the replication fork. This strand is synthesized in a continuous fashion.

ii. *Lagging Strand*—the daughter strand that is synthesized away from the replication fork. This strand is synthesized in a discontinuous fashion or in fragments called *Okazaki fragments*.

2. An enzyme called *primase* then synthesizes a short segment of *RNA* called a *primer* that is complementary to nucleotides on the DNA strand.

3. *DNA polymerases* bind to each separated strand and begin adding the proper complementary nucleotides to the primer to produce a new copy of each strand.

4. The bond formed between two nucleotides is a covalent bond between the deoxyribose sugar of one and the phosphate of the other.

5. Another enzyme, called *DNA ligase*, helps seal gaps between the many growing strands by taking a 5' phosphate and 3'

Replicates: 5' → 3'

* DNA polymerase only works

hydroxyl and linking them together, thereby helping join the Okazaki fragments into a single strand.

6. DNA polymerases keep moving along the strands until synthesis of both strands is completed.

D. Each new molecule of DNA consists of one of the original DNA strands hydrogen bonded to its newly synthesized complement.

E. *Proofreading* during DNA replication and repair of damaged DNA results in low mutation rates at the nucleotide level.

1. DNA polymerase makes mistakes at a rate of about 1/10,000 base pairs, but proofreading and repair mechanisms reduce that rate to 1/1,000,000,000.

2. Errors are usually corrected by enzymes that move along the new DNA molecule and replace any base that has been mismatched.

3. DNA molecules are also susceptible to damage by chemicals or radiation and are repaired in a similar manner.

Major Enzymes and Proteins in DNA Replication		
Enzyme	**Substrate**	**Action**
DNA helicase	Double-stranded DNA	Opens up the DNA strand for replication
Single-stranded binding proteins	Single-stranded DNA	Binds single-stranded DNA and keeps replication fork open
DNA primase	Single-stranded DNA	Lays down an RNA primer on single-stranded DNA for DNA polymerase to hook up with
DNA polymerase	Single-stranded DNA	Adds the complementary base to the daughter strand using the parental template. Follows base pairing rules; adenine with thymine, guanine with cytosine
DNA ligase	Single-stranded DNA	Links a 5' phosphate with a 3' hydroxyl on the lagging strand

RNA Structure and Gene Expression

I. Key Concepts

A. RNA transcription produces rRNA, tRNA, and mRNA, all of which have different roles during the process of translation.

B. In its sequence of nucleotides, mRNA carries the genetic information present in an organism's DNA (also in the form of sequences of nucleotides) from the nucleus to the ribosome.

C. Translation (protein synthesis) converts the information in the nucleotide sequences of mRNA into information in the form of the amino acid sequences of proteins that are critical to the structure and functioning of cells.

II. RNA Compared to DNA

A. *Ribonucleic acid (RNA)* structure is similar to DNA in some ways, but has very important differences.

1. RNA and DNA are both composed of nucleotides and both have strands consisting of a *sugar-phosphate backbone* with nitrogenous bases *attached to their sugars*.
2. RNA and DNA differ in the following significant ways:
    i. The 5-carbon sugar in RNA's backbone is *ribose* instead of deoxyribose.
    ii. Instead of thymine, RNA uses the pyrimidine base, *uracil*.
    iii. RNA does not form a stable double helix along its entire length with a complementary strand of RNA, as DNA does.

➤ RNA is often present in a *single-stranded state*, but is also can base pair with DNA, with itself, and with other RNA molecules.

➤ When it does form base pairs through hydrogen bonding, guanine pairs with cytosine, and adenine pairs with uracil.

iv. The two DNA strands in double-stranded DNA are antiparallel in directionality.

3. The following figure shows the complementarity of DNA and RNA:

```
-----G A T T C G T A A G G C------ D N A
-----C U A A G C A U U C C G------ R N A
```

B. The functions of RNA and DNA are different.

1. *DNA stores genetic information* on how to make RNA and proteins, and it passes this information from cell to cell and from parents to offspring.

2. Different types of RNA function in different ways.

i. *Messenger RNA (mRNA)* transfers genetic information from DNA in the nucleus to ribosomes in the cytoplasm, where its information is translated into proteins.

ii. *Ribosomal RNA (rRNA)* is incorporated into large complexes, called ribosomes, which are the sites of protein synthesis in the cytoplasm; rRNA also regulates gene expression at the level of mRNA transcription.

iii. *Transfer RNA (tRNA)* carries amino acids to a ribosome so they can be assembled into proteins.

3. The three types of RNA have different structures and functions, but all are made during the process of transcription, and all are important in the subsequent production of proteins during translation.

 III. **Transcription**

A. *Transcription*—the process whereby information contained in the nucleotide sequences of genes is transferred to RNA molecules.

B. Transcription has three basic steps and occurs in the nucleus of eukaryotes and the cytoplasm of prokaryotes:

1. *Initiation*—the initiation of transcription is controlled by interactions between various proteins and various regions of a gene.
   i. During initiation, the enzyme RNA polymerase, binds to the promoter and opens a portion of the gene to create a transcription bubble.
   ii. RNA polymerase begins to synthesize an RNA transcript complementary to only one strand of DNA called the template.
2. *Elongation*—during elongation, RNA polymerase moves along the template DNA, adding nucleotides to the elongating strand of RNA.
   i. RNA polymerase catalyzes the formation of a covalent bond between each new nucleotide by joining the ribose of one nucleotide to the phosphate of the other.
   ii. Each incoming nucleotide base pairs with its complementary pyrimidine or purine in the DNA template sequence on the gene.
3. *Termination*—when a termination sequence, or terminator, at the end of the gene is reached, RNA polymerase leaves the promoter, and the RNA transcript is released.

C. In eukaryotes, before RNA leaves the nucleus, it is modified.

D. rRNA—as part of the ribosomal subunits—plus mRNA and tRNA all travel through the nuclear pores of the nuclear envelope to the cytoplasm where they all participate in protein synthesis.

IV. Translation

A. *Translation*—also called protein synthesis, occurs in ribosomes in the cytoplasm of prokaryotic and eukaryotic cells.

B. Synthesis of protein from mRNA occurs in the 5′ to 3′ direction.

C. Requires all 3 RNA molecules (mRNA, tRNA, and rRNA) and codons (sets of 3 nucleotide RNA bases that code for amino acids).

D. The process of translation has three major steps:

1. *Initiation*—the start codon AUG (calls for the amino acid methionine) on the mRNA transcript is recognized by the ribosome.
   i. A tRNA—which carries the amino acid methionine and has the anticodon UAC—attaches to the ribosome at one of the sites set aside for tRNA on the ribosome, and its anticodon hydrogen bonds to the start codon, AUG.
2. *Elongation*—once the binding occurs, the entire ribosome translocates down another 3 bases and reads another codon sequence, where another tRNA brings in the appropriate amino acid.
   i. A peptide bond between the amino acids is formed via an enzymatic reaction promoted by the rRNA portion of the ribosome.
3. *Termination*—occurs when one of the stop codons (UAA, UGA, UAG) is read and the protein is released from the ribosome.

E. Protein activities can, in turn, affect the phenotype of an organism. Comparison of normal proteins with proteins that an abnormal allele is coding for allows scientists to begin to determine possible courses of treatments, if any. For example, in the case of albinism, the colorless compound DOPA is not converted to melanins. For an organism without albinism, the reaction should look like this:

DOPA— — — — — —·→ melanins

In albinism, the reaction looks like this:

DOPA— — — —//— —→ (no melanins)

Bottom line: A protein contained an abnormal allele that could not produce a specific product that was necessary for this reaction to proceed, and it resulted in albinism.

Nucleic Acid Technology and Applications

16

I. Key Concepts

A. DNA technology is a collection of procedures for manipulating and analyzing DNA that aid in all aspects of biological research and in developing technical applications for a wide range of purposes.

B. DNA technology has created many powerful tools for basic research, as well as for commercial use in agriculture and medicine.

II. Genetic Engineering Techniques

A. DNA Cloning

　　1. *Restriction Enzymes*—used to cut DNA molecules at specific locations called restriction sites.
　　　Example: Restriction site

```
        -----ACTGGA----                    -----A    CTGGA----
        -----TGACCT----      ────────►     -----TGACC    T----
```

　　2. *Recombinant DNA*—combining DNA sequences that would not normally occur together to form one piece of DNA. The enzyme DNA ligase is added to seal the strands together.

```
  -----A     CTGGA---- + [          ]  ──────►  -----A [          ] CTGGA----
  -----TGACC     T----   [          ] DNA Ligase -----TGACC [          ]  T----

     Cut DNA          Foreign DNA                    Recombinant DNA
```

3. Cloning Vector—original plasmid that is used to carry foreign DNA into a cell and replicate there.

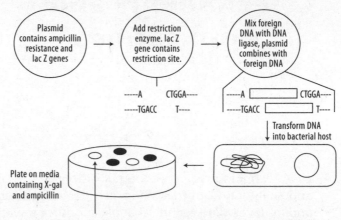

Bacterial clone that is white and grows on plate is clone containing recombinant DNA plasmid.

1. When the restriction enzyme is added to plasmid, *lac Z* is destroyed and nonfunctional. The *lac Z* gene produces the enzyme β-galactosidase, which breaks down the sugar X-gal causing the colony to appear blue. If the *lac Z* gene product is not made, the colony appears white; if the gene is functional, the colony appears blue.

2. Ampicillin resistance gene allows bacteria to grow in the presence of the antibiotic ampicillin.

3. Media is selective for clones that have the ampicillin resistance gene, and differential for blue or white colonies.

4. Colony that is growing on plate (ampicillin resistance) and white are correct clones carrying recombinant DNA.

B. DNA Gel Electrophoresis

1. DNA is placed in a gel made of a polysaccharide called agarose or acrylamide (used for smaller fragments).

2. Migration of DNA is based on size differential of DNA fragments. An electric field is passed through DNA molecules and the molecules travel toward the positive end (cathode) due to negative charge of phosphate on DNA.

3. Larger molecules travel slower; smaller molecules travel faster.
4. Marker DNA of a standard size is used to approximate the size of unknown molecules. Marker is measured in kilobase pairs.
5. Visualization of DNA is done by staining the gel with ethidium bromide, which increases the visual difference between DNA and the gel.

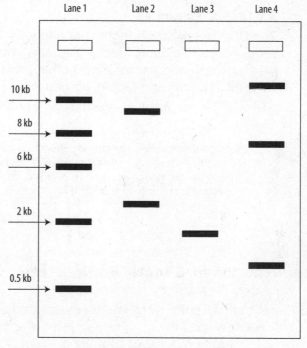

Lane 1 – Marker DNA/Standard Size
Lane 2 – 2 bands roughly 9 kb and 3 kb
Lane 3 – 1 band roughly 1.8 kb
Lane 4 – 3 bands roughly 12 kb, 7 kb, and 1 kb

Electrophoresis can be used for DNA and protein identification, isolation of different types of DNA or protein, calculating the size of fragments (DNA and protein), crime scene investigation, and genetic testing.

C. Polymerase Chain Reaction

   1. A method to take a small amount of DNA and amplify (increase) the amount.

   2. Based on progressive heating and cooling of DNA strands with the addition of primers and DNA polymerase.

D. DNA Fingerprinting

   1. A technique used by forensic scientists to help determine the DNA of individuals.

   2. The DNA of humans is highly homologous. There are sequences called *Short Tandem Repeats* (STRs). These repeats vary in length and size for each human, and therefore can be used as identifying factors of humans.

   3. STRs can be visualized using DNA gel electrophoresis.

Fully understanding the cloning process is considered a major concept in the AP Biology curriculum. You should understand how restriction enzymes and vectors are used in tandem to construct a recombinant plasmid.

III. Applications of Genetic Engineering

A. Genetic engineering techniques have created transgenic plants that are now used in agriculture to increase crop yields, reduce pesticide and fertilizer use, improve nutritional quality of grains, and create plants tolerant to extreme weather conditions such as drought.

B. Practical uses of DNA technology in medicine include production of vaccines and other pharmaceutical products.

   1. Genetic analysis and transgenic organisms are used to create more effective vaccines that are less likely to cause disease than traditionally manufactured vaccines.

2. Cloning human genes in bacteria using expression vectors has resulted in large supplies of important medicines such as insulin to treat diabetes and interferons and interleukins to treat acquired immunodeficiency syndrome (AIDS).

C. Research applications of genetic engineering have been extensive. For example, cloning, RFLP analysis, PCR and related chromosome mapping techniques have been used to map the entire human genome (as well as the genomes of many other organisms critical to basic research), creating computer databases that are widely available to researchers in all fields, including medicine, mathematics, engineering, computer technology, and other biology disciplines.

The Cell Cycle and Mitosis

I. Key Concepts

A. The number of chromosomes an organism has in its body cells does not vary from cell to cell, nor does it vary from organism to organism of the same species.

B. The genetic material (DNA) of all organisms is contained on chromosomes that become especially compact during cell reproduction.

C. The cell cycle is the life cycle of a cell and includes time periods when a cell is not dividing as well as those when it undergoes cell division.

D. *Sister chromatids* consist of two duplicated chromosomes held together at the centromere.

E. *Mitosis* occurs in eukaryotes and produces cells with nearly identical genetic makeup.

   1. Mitosis is used for the purpose of organismal reproduction in single-celled organisms.
   2. It is used for purposes of development and cell replacement in the normal growth and maintenance of the bodies of multicellular organisms.

F. Prokaryotes generally reproduce by a process called *binary fission*.

II. The Cell Cycle

A. The two main phases of the cell cycle are interphase and cell division.

B. *Interphase*—Cells spend most of their time in interphase.

"Cell being a cell"

1. G_1 *phase*—The first phase of interphase is G_1 during which the new cell grows to mature size and may begin to carry out its specific function.

2. *S phase*—If the cell is going to divide again, it duplicates its chromosomes during the S phase by the process of DNA replication.

Chpt 14.

3. G_2 *phase*—Once the DNA is replicated, the cell enters G_2 during which it prepares for cell division.

4. G_0 *phase*—Some cells do not divide, or they delay division; these cells enter the G_0 phase sometime during G_1.

DNA = Chromatin

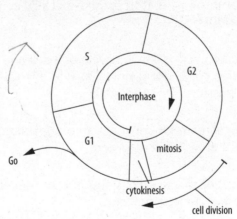

C. The second main phase of the cell cycle is cell division, which includes mitosis and cytokinesis.

III. Mitosis

A. *Mitosis*—the division of the nucleus, which leads to the separation of the chromosomes that were previously duplicated in the S phase to produce two chromatids that are attached at their centromeres; it has four main stages.

B. Phases of Mitosis

1. Prophase—
   i. The chromatids <u>condense</u>.
   ii. The nuclear membrane <u>breaks down and disappears</u>.
   iii. A cytoskeletal structure called the *mitotic spindle* forms, and is used to pull the chromatids apart to either pole of the cell.
      ➤ Two *centrosomes* are synthesized.
      ➤ In animal cells, the centrosomes contain small cylindrical bodies called *centrioles*.
      ➤ The centrosomes move to opposite sides of the cell.
      ➤ Spindle fibers, made of microtubules, radiate outward from the centrosomes and some, called *kinetochore fibers*, attach to each chromosome at its centromere.

2. *Metaphase*—During metaphase, <u>the chromatids line up across the center of the cell</u> called the *metaphase plate*.

3. *Anaphase*—<u>Chromatids</u> separate from one another during anaphase, at which point each chromatid is now considered to be an <u>individual chromosome</u>.
   i. The centromeres of the chromatids split.
   ii. The kinetochore fibers pull one copy of each chromosome to one pole and the rest to the other side of the cell.

4. *Telophase*—the last stage of mitosis.
   i. The mitotic spindle disassembles.
   ii. <u>The chromosomes unwind from their highly compacted</u> state.
   iii. A new nuclear membrane forms and surrounds each new complete set of chromosomes.

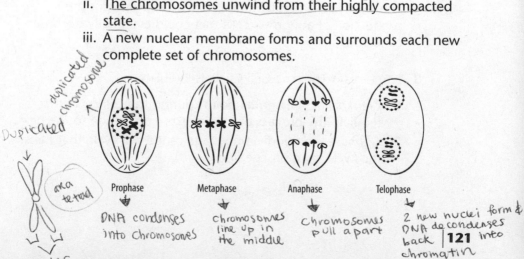

duplicated chromosome

Duplicated

aka tetrad

Sister Chromatid

Prophase
DNA condenses into chromosomes

Metaphase
Chromosomes line up in the middle

Anaphase
Chromosomes pull apart

Telophase
2 new nuclei form & DNA de condenses back into chromatin

121

IV. Cytokinesis

A. *Cytokinesis*—the division of the cytoplasm following mitosis whereby the two newly formed nuclei become incorporated into separate cells.

  1. In animal cells, a special collection of microfilaments of the cytoskeleton form a *cleavage furrow* in the center of the cell, which causes the cell to be pinched into two cells.

  2. In plant cells, vesicles from the Golgi apparatus form a *cell plate* in the center of the cell along which new cell wall material is deposited between the two newly forming plasma membranes.

Cytokinesis in Plant and Animal Cells

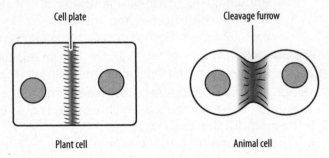

Plant cell   Animal cell

V. Regulation of the Cell Cycle

A. The cell cycle is regulated in order to help prevent the production of abnormal cells that could eventually become cancerous.

B. Some Ways That the Cell Cycle Is Regulated

  1. *Checkpoints*—essential points during the cell cycle that regulate the process of passing from one stage to the next.

  2. *G_o phase*—a nondividing stage of the cell cycle that halts the cycle from proceeding.

3. *Growth factor*—protein/hormone that promotes the division of cells.
4. *Density-dependent Inhibition*—process in which cells stop dividing when they are in contact with each other.
5. *Anchorage dependence*—cells must be attached to something in order to divide properly.

C. When Regulation of the Cell Cycle Does Not Work

1. *Cancer Cell*—cells that are said to be "transformed" from normal cells to cancer cells and do not exhibit density-dependent inhibition. Have uncontrolled growth pattern.
2. *Tumor*—a pocket of abnormal cells among normal cells.
3. *Benign tumor*—nonspreading of abnormal cells.
4. *Malignant tumor*—abnormal cells that invade and impact the normal function of an organ.
5. *Metastasis*—spreading of malignant tumor to other parts of the body.

Meiosis

I. Asexual and Sexual Reproduction

A. *Asexual Reproduction*—a form of reproduction not requiring meiosis or fertilization; only passes a copy of genes to its progeny. It's a type of reproduction in which there is no variation in genetic makeup. Bacteria reproduce via asexual reproduction.

B. *Clone*—an individual that arises from asexual reproduction.

C. *Sexual Reproduction*—a type of reproduction that involves variation because two parents give rise to their progeny.

Major evolutionary advantage because of genetic variation.

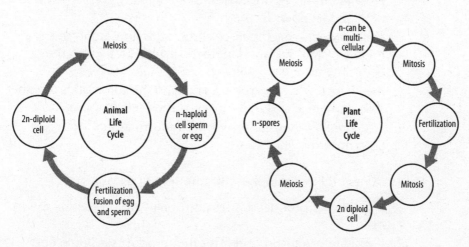

II. Meiosis

A. Meiosis, like mitosis, is preceded by replication of chromosomes.

B. *Meiosis I* is the division where homologous pairs of chromosomes are separated from one another into two cells that are haploid, and it can divided into four stages:

1. *Prophase I*—Tetrads are formed or the pairing of homologous chromosomes via *synapsis*; *chiasmata* or the site of crossing over/exchange of genetic material is formed during this phase.
2. *Metaphase I*—homologous chromosomes pair with each other at the metaphase plate.
3. *Anaphase I*—homologous chromosomes separate and sister chromatids stay together.
4. *Telophase I*—the movement of chromosomes to the poles is completed.

C. *Meiosis II* occurs in each of the two new cells after meiosis I is completed. No DNA replication occurs between meiosis I and meiosis II. Meiosis II has four stages:

1. *Prophase II*—the chromosomes are already condensed and new spindle fibers form.
2. *Metaphase II*—each pair of chromatids lines up in the middle of the cell, and kinetochore fibers attach to the centromeres of each pair.
3. *Anaphase II*—the centromeres holding the chromatids together split, and one chromatid moves to each side of the cell.
4. *Telophase II*—nuclear envelopes re-form around the chromosomes.

D. *Cytokinesis* follows telophase II, resulting in gametes.

1. If sperm are produced, meiosis usually produces four sperm cells.
2. If ova are produced, meiosis often produces a single egg cell, while the other three cells die or have other functions in reproduction.

Every AP Biology test-taker should know the main differences between meiosis and mitosis. Mitosis produces diploid identical cells that have no genetic variation. Meiosis produces gametes (haploid) that are genetically different because of crossing over in Prophase I of meiosis. Similarly, know the stages of mitosis and meiosis and special structures that are formed.

Meiosis I and Meiosis II

Early prophase I

Late Prophase I (synapsis begins)

Later prophase I (synapsis has occurred)

Later prophase (cross-over occurring)

Anaphase I

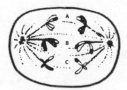

Telophase I

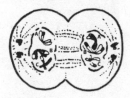

Prophase II

Metaphase II

Anaphase II

After telophase II

III. **Comparison of Mitosis and Meiosis**

Event	Mitosis	Meiosis
DNA Replication	Occurs during interphase	Occurs during interphase
Homologous Chromosomes	Align one after another on metaphase plate	Pair with each other during metaphase I. Align one after another on metaphase plate during metaphase II
Sister Chromatid Separation	Anaphase	Meiosis II Anaphase II
Divisions	1	2
Cells Produced	2 Diploid—genetically identical	4 Haploid—genetically different
Crossing Over	Does not occur	Meiosis I Prophase I

Structure and Inheritance of Chromosomes

I. Key Concepts

A. In the mid-1800s, without knowledge of chromosomes or genes, but with careful experimentation and statistical analysis, Gregor Mendel worked out the basic rules of *heredity*—the inheritance of characteristics from generation to generation—that occur during sexual reproduction.

B. The *Law of Segregation* is the principle that the two determinants of a characteristic, called *alleles* of a gene, are separated during meiosis and are distributed to separate gametes.

C. The *Law of Independent Assortment* is the principle that the segregation of one set of alleles into gametes, which determine one characteristic, is independent of the segregation of a second set of alleles governing a second characteristic.

D. Genes are located on chromosomes, and in sexually reproducing organisms, alleles—different versions of the same gene—are located at the same position on homologous chromosomes.

E. Understanding the use of probabilities and Punnett squares is important for determining genetic cross outcomes.

F. The sex chromosomes, X and Y, determine the sex of mammals and many insects. Genes on the X and Y chromosomes have special patterns of inheritance called X-linkage and Y-linkage.

Genes on the same chromosome do not assort independently if they are linked; linkage can be used to map the relative locations of genes on a chromosome.

H. Pedigree analysis is used to study the inheritance of human genes.

II. Chromosomal Structure

A. Structure and Function of Eukaryotic Chromosomes

Part	Structure	Function
Genes	Made up of the nucleic acid DNA	Will be transcribed onto mRNA Will be translated for proteins
Chromatids	Two replicated chromosomes that are held together at the centromere	Allows proper segregation of chromosome during meiosis and mitosis
Centromere	DNA region found near the middle (not always) chromosome	Hold chromatids together to form a chromosome
Chromatin	DNA and protein combination	Aids in packaging DNA, DNA replication, and expression of proteins
Kinetochore	Proteins	Allows for the attachment of the mitotic spindle to the centromere
Nucleosomes	Histone proteins and DNA	Aids in packaging of DNA
Telomeres	Ends of DNA	Protection against the destruction of the DNA from nucleases

Remember: Eukaryotic DNA is linear, meaning it has definite ends.

Most eukaryotic organisms are diploid. Fungi, such as yeast, can exist as haploid or diploid.

B. Structure and Function of Prokaryotic Chromosome as a Comparison

1. Circular in shape and much smaller than eukaryotic chromosome.
2. Genes are arranged in *operons*—one promoter controlling many genes.
3. Transcription and translation are coupled processes.
4. *Plasmids* are prevalent—extra chromosomal pieces of DNA that carry antibiotic resistance. They are not part of the chromosome. Autonomously replicating.
5. One origin of replication.
6. No histone proteins to condense, but DNA is supercoiled.

III. Inheritance Patterns

A. Terms

1. *Characteristic*—an inheritable feature such as hair color (phenotype).
2. *Trait*—a variant of a characteristic. Example: red or blond hair color.
3. *Allele*—alternative form of a gene, such as tall (*T*) plants are dominant to short plants (*t*).
4. *Dominant allele*—the allele that is fully expressed.
5. *Recessive allele*—the allele that is not expressed.
6. *Genotype*—the genetic makeup of an organism.
7. *Phenotype*—organism's appearance.

B. *Law of Segregation*—is observed with monohybrid crosses or crosses for a single characteristic. The law states that each trait must result from two distinct factors and that these factors separate from each other during reproduction and are incorporated into separate gametes.

C. *Law of Independent Assortment*—is observed with dihybrid crosses or crossed between two different characters. The law states that alleles assort independently from each other; therefore, dominant alleles can combine with recessive alleles.

D. Genetic Crosses

1. *Monohybrid cross*—a cross that tracks the inheritance pattern of a single character. Apply the Law of Segregation.

   Example: In pea plants, tall (T) plants are dominant to short plants (t). There are three allelic combinations:
   - ➤ TT—homozygous dominant (true breeding)/Tall
   - ➤ Tt—heterozygous or hybrid/Tall
   - ➤ Tt—homozygous recessive (true breeding)/Short

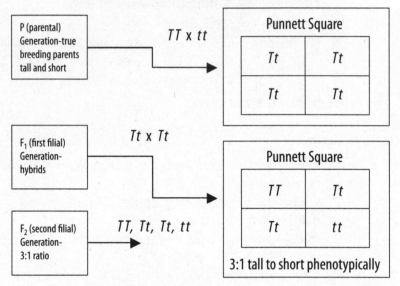

Cross two true breeding plants of tall and short

P (parental) Generation-true breeding parents tall and short	TT x tt	Punnett Square	
		Tt	Tt
		Tt	Tt

F₁ (first filial) Generation-hybrids	Tt x Tt	Punnett Square	
		TT	Tt
F₂ (second filial) Generation-3:1 ratio	TT, Tt, Tt, tt	Tt	tt
		3:1 tall to short phenotypically	

- ➤ Note: Punnett squares can be used to determine the genotypes and phenotypes of progeny from a genetic cross. Monohybrid or multi-hybrid crosses can be used.

2. *Test cross*—A cross that determines whether the dominant parent is homozygous dominant or heterozygous. Always cross the dominant parent to a homozygous recessive. Assume black (B) is dominant to white (b) for cat coat color.

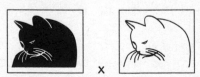

x

Black parent could be BB or Bb. White parent is bb.
➤ If BB x bb, all progeny will be black carriers.

➤ If Bb x bb, ½ of the progeny are black and ½ are white.

3. Dihybrid Cross—a cross between two different characteristics; demonstrates Law of Independent Assortment.

Example: In pea plants, tall (T) plants are dominant to short plants (t). Green leaf (G) is dominant to yellow leaf (g).

Cross two true breeding plants of tall green and short yellow

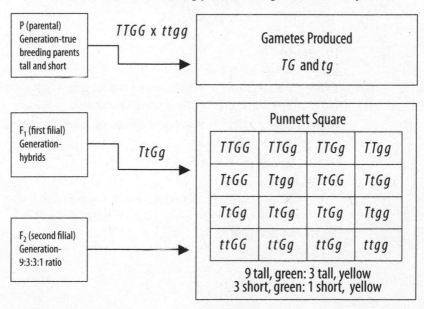

P (parental) Generation-true breeding parents tall and short	*TTGG* x *ttgg*	Gametes Produced *TG* and *tg*		

| F₁ (first filial) Generation-hybrids | *TtGg* | Punnett Square | | |

| F₂ (second filial) Generation- 9:3:3:1 ratio | | | | |

TTGG	*TTGg*	*TTGg*	*TTgg*
TtGG	*Ttgg*	*TtGG*	*TtGg*
TtGg	*TtGg*	*TtGg*	*Ttgg*
ttGG	*ttGg*	*ttGg*	*ttgg*

9 tall, green: 3 tall, yellow
3 short, green: 1 short, yellow

E. Using the Laws of Probability in Genetics

1. *Probability (p)*—the number of times an event is expected to occur divided by the number of opportunities for the event to occur. A probability can be expressed as a fraction, a percentage, or a decimal; for example:
   i. ¼ = 0.25 = 25%
   ii. ½ = 0.5 = 50%
   iii. 1/1 = 1 = 100%

2. *Law of Multiplication*—used to calculate the probability of independent events occurring; therefore, for genes that are linked the law of multiplication cannot be followed.

   Example 1: Assume the following cross: AaBbCc x AabbCC. What are the chances of the following progeny?
   (a) AabbCC
   (b) aabbCc
   (c) AAbbCC

   Answer: Perform each individual monohybrid cross and use the law of multiplication.

   Aa x Aa = 1/2 Aa, 1/4 aa, 1/4 AA
   Bb x bb = 1/2 Bb, 1/2 bb
   Cc x CC = 1/2 Cc, 1/2 CC
   (a) AabbCC = ½ x ½ x ½ = 1/8
   (b) aabbCc = ¼ x ½ x ½ = 1/16
   (c) AAbbCC = ¼ x ½ x ½ = 1/16

   Example 2: Assume the following genotype: AaBBCcddEeFf. How many different gametes are possible?

   Answer: Determine how many different gametes are possible for each set of alleles.

   Aa = 2 (either A or a)
   BB = 1 (only B)
   Cc = 2 (either C or c)   2 x 1 x 2 x 1 x 2 x 2 =
   dd = 1 (only d)   16 different gametes
   Ee = 2 (either E or e)
   Ff = 2 (either F or f)

F. *Non-Mendelian Genetics*—genetics that do not follow the inheritance patterns of Mendel's initial pea plant experiments.

1. *Incomplete dominance*—the phenotype of the offspring has an appearance that is between that of both parents. This is not a blending hypothesis. The dominant allele is not fully expressed.

Snapdragons

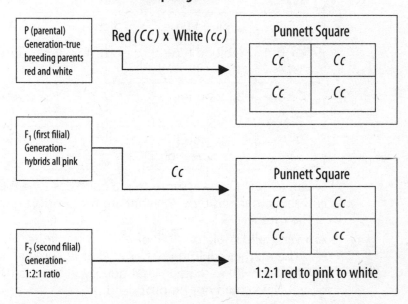

2. *Codominance*—both alleles are expressed at the same time.
    i. MN Blood system (M and N are blood group antigens found on the cell surface of a red blood cell).
    ii. There are three allelic combinations:
        ➤ *MM*—homozygous dominant (only produce M antigen on cell surface).
        ➤ *MN*—heterozygous (produce M and N antigens on cell surface).
        ➤ *NN*—homozygous recessive (only produce N antigen on cell surface).

3. *Multiple Alleles*—many different alleles can control the expression of a character.
   i. *ABO Blood System*—carbohydrate antigens found on the cell surface.

Geno-type	Phenotype	Antigen on Cell Surface of Red Blood Cell	Antibodies Present in Blood
ii	O blood type	None	Anti A, Anti B
I^A I^A or *I^A i*	A blood type	B	Anti A
I^B I^B or *I^B i*	B blood type	A	Anti B
I^A I^B	AB blood type	AB	None

Example ABO Cross: Assume that a child has type B blood and the father was type A. What are the possible genotypes of the mother?

Answer: Child could be I^B I^B or I^B i and the father could be I^A i. The mother could be either I^B I^B or I^B i or I^A I^B. If the father was I^A I^A no matter what genotype the mother is, a type B child could not be produced.

4. *Pleiotropy*—one gene causes multiple different phenotypic effects on an organism.
   ➤ An example of pleiotropy is PKU, which causes the following:

Human Disease PKU (phenylketouria) Phenotypes		
Mental Retardation	Hair Loss	Skin Pigmentation

5. *Epistasis*—one gene affecting the expression of another gene. F_2 offspring phenotypic ratio is usually 9:3:4.
6. *Polygenic inheritance*—two or more genes affecting one phenotype. Examples include skin color and cancer, which

is the most common polygenic inherited disorder. Polygenic inheritance leads to a bell curve distribution of phenotypes.

G. *Pedigree Analysis*—a visual depiction of inheritance patterns in multiple family generations.

1. Basic Rules
   i. If two affected people have an unaffected child, it must be a dominant pedigree: [A] is the dominant mutant allele and [a] is the recessive allele. Both parents are Aa (hybrid carriers) and the unaffected child is aa.
   ii. If two unaffected people have an affected child, it is a recessive pedigree: [A] is the dominant allele and [a] is the recessive allele. Both parents are Aa (hybrid carriers) and the affected child is aa.
   iii. If every affected person has an affected parent, it is a dominant pedigree (no skipping of generations).
   iv. Dominant traits never skip generations, while recessive traits can skip.
   v. Squares are male.
   vi. Circles are females.
   vii. Mating is indicated by the connection with a line.

   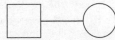

   viii. Filled-in circles or squares indicate affected person.

   ix. *Sex-linked dominant*—all females descending from the affected males have the disease.
   x. *Sex-linked recessive*—no male carriers possible and skips generations.
   xi. *Autosomal recessive*—carriers are present, so skips generations. 50% males and females affected.
   xii. *Autosomal dominant*—no carriers or skipping of generations. 50% males and females affected.

H. Autosomal Genetic Disorders

Recessive Inherited Disorder—absence or malfunction of protein Must receive both nonfunctional copies from parents; therefore, affected individual is homozygous recessive (aa).	
Disease	**Outcome**
Albinism	Lack of pigment in the skin, eyes, and hair. May lead to skin cancers.
Cystic Fibrosis	Defective or absent chloride channel protein in membranes, causing a build-up of mucus in lungs. Person is prone to bacterial infections.
Tay-Sachs	Defective or absent lipase enzyme in brain. Predominant in Jewish population.
Sickle cell disease	Defective hemoglobin protein. Mostly affects the African-American population.

Dominant Inherited Disorder—absence or malfunction of protein Must receive at least one nonfunctional copy from one parent; therefore, affected individual is heterozygous (Aa) or homozygous dominant (AA).	
Disease	**Outcome**
Achondroplasia	Dwarfism
Huntington's Disease	Degenerative breakdown of the nervous system.

1. As researchers gain more knowledge about many of these genetic disorders, there are also numerous social, medical, and ethical issues surrounding these disorders. Tay-Sachs disease, for example, can lead to pre-conception screenings to determine the probability of the couple having a child with this disorder; it can also lead to social, medical, and ethical challenges as how to approach a pregnancy of a child diagnosed with Tay-Sachs.

2. There is also a civic issue surrounding genetic disorders. As companies begin to unveil the genetic makeup of many of these disorders and develop tests and treatments for them, who or what owns this knowledge? Can a gene or a test for a gene be patented? Do insurance companies have "rights" to prescreenings done by individuals who are at higher risks for carrying certain genes, like those that are linked to breast cancer?

I. Linked Genes—the alleles of two genes located on the same chromosome often do not show independent assortment; instead, they exhibit a special type of inheritance called *linkage*.

1. If two genes are located close together on the same chromosome, they do not assort independently because they are physically linked to each other.
2. *Example of Linked Genes Experiment*—Thomas Hunt Morgan performed genetic crosses with the fruit fly (*Drosophila melanogaster*) and used the following terms:
   i. *Wild type*—most common phenotype in the population.
   ii. *Mutants*—alternative phenotypes to the wild type.
3. Morgan performed the following dihybrid mating:
   Example: In fruit flies, gray (g+) body color is dominant to black body color (g). Normal wings (w+) are dominant to dumpy wings (w).
   Cross a double heterozygote to a double recessive (g+g w+w x ggww).
   Expected phenotypes of 1,000 offspring would be:
       250 wild type (gray normal)/parental phenotype
       250 black dumpy/parental phenotype
       250 gray dumpy/recombinant phenotype
       250 black normal/recombinant phenotype
   Observed phenotypes of 1,000 offspring were:
       450 wild type (gray normal) /parental phenotype
       450 black dumpy/parental phenotype
       50 gray dumpy/recombinant phenotype
       50 black normal/recombinant phenotype
   The high number of observed parental phenotypes indicated that the genes for body color and wings were linked to each other. Linked genes are on the same chromosome and are very close to each other. Linked

genes are inherited together and recombination between the genes is very low.

Calculation of Recombination Frequency or the measure of genetic linkage between 2 genes (also called map units).

$$\text{Recombination Frequency} = \frac{\text{\# of recombinants}}{\text{total offspring}} \times 100$$

Using the data above:

$$\text{Recombination Frequency} = \frac{100}{1000} \times 100 = 10\%$$

Only 10% of the time will there be recombination between the genes for body type and wings.

4. *Genetic Maps*
   i. Recombination Frequency allows you to create genetic maps that estimate the distance between genes.

   Example: Assume the following Recombination Frequencies. Determine the genetic map for genes W, X, Y, Z.

   W-Y, 7 map units
   W-X, 26 map units
   W-Z, 24 map units
   Y-X, 19 map units
   Y-Z, 31 map units

   Answer: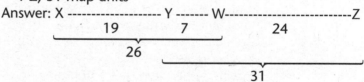

5. *Sex-linked Genes*—genes that are carried on the X-chromosome.
   i. Females carry two X chromosomes, XX.
   ii. Males carry 1X and 1Y, XY.
   iii. Inheritance patterns of sex-linked genes:
   ➤ A father will always transmit the sex-linked trait to his daughter. His son receives the Y, and does not inherit the trait.
   ➤ Only females can be carriers of sex-linked traits. Therefore, a carrier female who mates with a normal male transmits the mutant allele to half her sons

and half her daughters. Examples of sex-linked traits include *hemophilia* and *muscular dystrophy*.

➤ *Barr body*—one of the female's X chromosomes is randomly inactivated in order to have the same gene dosage as males for sex chromosomes. The chromosome tends to look smaller in physical structure. Example of the phenotypic output of X-inactivation are *calico-colored cats*.

Practicing a variety of genetics problems is an essential test preparation activity. Some genetics problems provide information about parents and ask about phenotypic ratios of their progeny; others give information about progeny and ask about the parents' genotypes or phenotypes. Be prepared for both.

Regulation of Gene Expression

 I. **Key Concepts**

A. *Gene expression*—the transcription and translation of a gene into protein—is controlled by DNA sequences surrounding the coding region of a gene and by regulatory proteins that bind to these sequences.

B. Control over gene expression is important for determining when, and in which cells, a protein will be made.

C. Gene regulation in bacteria involves control over the transcription of operons, and in eukaryotes involves multiple levels before, during, and after transcription and translation.

D. *Regulatory proteins* provide both positive and negative control mechanisms for gene expression. They inhibit gene expression by binding to DNA and blocking transcription (*negative control*), and they can stimulate gene expression by binding to DNA and stimulating transcription (*positive control*) or binding to repressors to inactivate their repressor functions.

E. Some genes, like *ribosomal genes*, are always turned "on" and are continuously expressed.

II. **Gene Regulation in Prokaryotes**

A. In *prokaryotes*, such as *bacteria*, the single chromosome contains many genes that are organized into operons.

1. An *operon* contains a promoter, an operator, and a group of structural genes.
   i. Several *structural genes*, often coding for proteins involved in the same metabolic process, are under the control of a promoter and operator.
   ii. A *promoter* is the part of the operon to which RNA polymerase binds in order to begin transcribing the structural genes.
   iii. An *operator* is the part of the operon to which a repressor protein can bind in order to stop expression of the structural genes.
2. *Repression*—occurs when a regulatory protein, called a *repressor*, binds to the operator, thereby blocking RNA polymerase from transcribing the genes.
3. *Induction*—occurs when a substance, called an *inducer*, binds to the repressor protein, inactivates it, and keeps it from binding to the operator, thereby activating transcription of the genes.

Regulation of a Bacterial Operon: The Lac Operon

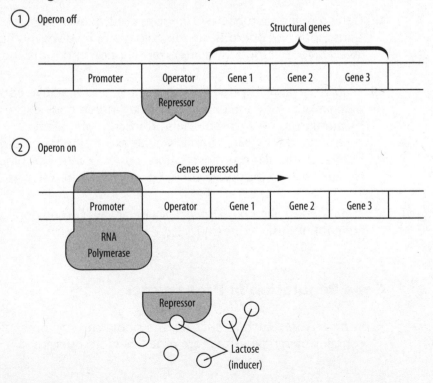

4. The *lac operon* is induced by the sugar, lactose.

    i.   The structural genes of the *lac operon* control the utilization of lactose by the bacteria, *Escherichia coli* (*E. coli*).

    ii. When lactose is not present in its environment, *E. coli* has no need to turn on the *lac* operon because there is no lactose to metabolize; so the repressor is bound to the operator, and the genes are turned off.

    iii. If lactose becomes available, it acts as an inducer by binding to the repressor and inactivating it so that RNA polymerase can bind to the promoter and transcribe the structural genes.

    iv. When the structural genes are translated into proteins, the proteins help the bacteria use lactose as an energy source.

    v. As the lactose is broken down and used by the bacteria, eventually none is left to bind to the repressor to keep it inactive; so the repressor binds once again to the operator, thereby turning off expression of the structural genes.

5. In this way, bacteria save resources and energy by turning off an operon when its gene products are not needed.

Test Tip

Know the basic parts of an operon and how operons are turned on and off.

III. Gene Regulation in Eukaryotes

A. *Eukaryotic gene expression* is controlled at many levels as DNA's information is converted to mRNA and then into protein.

Opportunities for Control over Eukaryotic Gene Expression

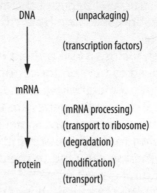

DNA   (unpackaging)

(transcription factors)

mRNA

(mRNA processing)
(transport to ribosome)
(degradation)

Protein   (modification)
(transport)

1. Only a small fraction of the genes in the genomes of multicellular organisms need to be expressed in any given cell type at any given time.
2. Eukaryotic genes are not organized into operons as in bacteria.
   i. Each gene has its own *promoter* and other *control elements* present in the DNA sequences surrounding the gene.
   ii. Many types of *regulatory proteins* bind to these elements to determine the timing and level of expression of any particular gene.
3. Controls over gene expression can occur before, during, or after transcription and translation.
   i. *Before transcription* can occur, the region of the genome in which the gene is located needs to be *unpackaged* to allow access to regulatory proteins.
      ➤ During interphase, some DNA remains as tightly packed as it was during mitosis and is called *heterochromatin*; genes in these regions of the genome are not able to be expressed.
      ➤ The uncoiling of DNA in a region—which is partly controlled by *histone proteins* associated with DNA—produces less tightly packed *euchromatin*, allowing regulatory proteins to access genes in that region.
   ii. To allow for *transcription* after a region of DNA has been unpackaged, regulatory proteins, such as *transcription*

factors, bind to the regulatory elements of the gene, thereby allowing access to the promoter by RNA polymerase.

> ➤ Some transcription factors bind directly to the promoter to assist the binding of RNA polymerase.
> ➤ Other transcription factors bind to DNA regions surrounding the gene, called *enhancers*, to further assist RNA polymerase binding.
> ➤ Once *RNA polymerase* binds to the promoter, transcription can occur.
> ➤ By controlling the synthesis and activity of transcription factors, a cell determines which gene will be expressed and when.

iii. *After transcription*, opportunities for control of gene expression can occur during *RNA processing*, which may involve the speed or types of splicing that occur to remove introns and join exons together to produce the mature mRNA.

iv. *Before translation*, the amount of mRNA sent to the ribosome can be controlled by how efficiently it is *transported from* the *nucleus* to the ribosome, or by how many *mRNA transcripts* are *degraded* along the way.

v. After translation, opportunities for further regulation of gene expression may involve *protein modifications* and *transport*, such as clipping off small parts of the protein, adding sugars to the protein, or reading signals located in the protein's amino acid sequence that determine where the protein is to be transported.

> ➤ Gene regulation in both prokaryotes and eukaryotes accounts for some of the phenotypic diversity of organisms, even though they may have similar genes.

Genetic Variation

I. Key Concepts

A. Many different types of mutations of nucleotides and chromosomes can occur during DNA synthesis and meiosis, resulting in different effects.

B. Mutations in somatic cells and germ cells can lead to the formation of cancer.

II. Mutations: Causes, Types, and Consequences

A. *Mutations during DNA Replication*—mutations of one or a few nucleotides in DNA are called point mutations and usually occur during DNA replication.

  1. *Substitution mutations*—occur when a nucleotide has been altered or incorrectly paired during DNA synthesis, thereby changing it to another nucleotide.
     i. Substitution that occurs within a *coding region* of a gene may or may not cause a change in the amino acid in that position of the protein.
     ii. Some substitutions can result in *detrimental effects*, such as changes in blood protein hemoglobin that result in sickle cell anemia disease
  2. *Insertion or deletion point mutations*—occur when one or a few nucleotides are inserted or deleted.
     i. Insertions or deletion of groups of three nucleotides within a coding region of a gene may simply cause insertion or deletion of amino acids in a protein.

ii. Insertion or deletion of one nucleotide, or groups of nucleotides that are not divisible by three, result in *frameshift mutations*—causing the reading frame (the coding region) to shift; this mutation causes all the amino acids after the site of the mutation to also be altered.

iii. A frameshift mutation, especially near the start of the gene, almost always results in a completely defective protein.

B. *Mutations during Meiosis*—mutations occurring during meiosis (as opposed to those that occur during DNA replication) can involve parts of chromosomes or whole chromosomes.

1. *Mutations in chromosome structure*—where regions of DNA much larger than those involved in point mutations are involved—are due to chromosome breakage.

   i. *Deletion*—occurs if a region of the chromosome (that does not contain a centromere) is broken and does not rejoin the chromosome.

   ii. *Duplication*—occurs if a broken portion of a chromosome becomes incorporated into its homologous chromosome.

   iii. *Inversion*—occurs if the broken portion of the chromosome may also be inverted and reattached to the same chromosome.

   iv. *Translocation*—occurs if a portion of the chromosome is moved from one chromosome to a chromosome that is not its homologue.

2. *Mutations in chromosome number*—involve the loss or gain of whole chromosomes, or duplication of whole genomes (all of an organism's chromosomes).

   i. *Nondisjunction*—the failure of chromosomes to separate properly during meiosis—can result in the production of cells with abnormal numbers of chromosomes.

      ➤ Nondisjunction can occur during meiosis I or meiosis II. When nondisjunction occurs during meiosis I, a complete tetrad of one pair of homologous chromosomes moves to one side of the cell. When nondisjunction occurs during meiosis II, a pair of sister chromatids fails to separate, pulling both chromatids to one side of the cell.

➤ Regardless of whether nondisjunction occurs during meiosis I or meiosis II, the result is that some gametes are missing one chromosome and others have two copies of that chromosome.

➤ *A normal gamete and a gamete missing one chromosome (n − 1) join to create a zygote with a missing chromosome that is called a monosomic.*

➤ *A normal gamete and a gamete with an extra chromosome (n + 1) may join to create a zygote with an extra chromosome that is called a trisomic.*

C. *Consequences of Mutations*—the effects of mutations run the gamut from neutral to lethal, but mutations also provide the raw material for evolution by natural selection.

1. Mutations can occur in somatic cells or germ cells.
   i. If a mutation occurs in a germ cell, it can be inherited because the products of germ cells are gametes.
   ii. If a mutation occurs in somatic cells, it is not inherited through sexual reproduction, but it may have other effects in the organism in which it occurs, such as the development of *cancer*.

2. Germ cell mutations that do not affect the fitness of an organism are known as *neutral mutations*.
   i. Neutral mutations result in neutral variation—variations between organisms that do not seem to affect evolutionary fitness.
   ii. Many, if not most, of the DNA differences between members of the same species, such as those revealed by DNA fingerprinting, may not affect an organism's fitness.
   iii. Neutral variation may be important during evolution because environmental conditions vary over time, and a variation that is neutral under one set of conditions may be beneficial under a different set of conditions.
   iv. Sometimes the variation can improve the evolutionary fitness of the individual under a certain set of conditions; for example, *sickle cell anemia* allows those with this condition to be less likely to contract malaria, and in parts of the world in which malaria is rampant, avoiding contracting malaria improves one's evolutionary fitness under those conditions.

3. *Polyploidy*—occurs in plants, can create *new species* in one or a few generations.
4. *Allelic variation*—germ cell mutations create the *allelic variation* that underlies *phenotypic variation*.
   i. A mutation in the regulatory parts of genes, such as a promoter, can affect when, where, and how much of a protein is produced in different cells in a body.
   ii. A mutation in a gene is important as a regulatory component in processes such as metabolism, cell to cell communication, growth, development, etc., and can create significant differences between individuals and between species.
   iii. Mutations that alter the expression of proteins or their amino acid sequences can result in *lack of a protein* or in a *nonfunctional protein* that may cause *inherited diseases*.
5. *Lethal Mutations*—some germ cell mutations can include point mutations or the loss of parts of chromosomes or whole chromosomes, and result in the death of an organism before birth.
6. *Cancer*—a combination of germ cell and somatic cell mutations are involved in the development of cancer.
   i. *Tumor cells* undergo cell division much more often than normal cells because they lack control over the cell cycle and other cell growth processes.
      ➤ Cells normally undergo cell division during activities such as development, growth, maintenance, and repair of an organism's body.
      ➤ Some cells divide repeatedly to produce masses of cells called *tumors*.
      ➤ Tumors become dangerous if they interfere with normal body functions, and if they spread—*metastasize*—to multiple locations within the body, causing *cancer*.
   ii. Mutations that cause tumors can occur in germ cells or somatic cells.
      ➤ A mutation in a germ cell can be inherited, and people with inherited mutations are more susceptible to developing some types of cancer.

➤ Mutations in somatic cells add to the effects of inherited mutations, making the development of cancer more likely.

iii. There are several *types of genes* that, when mutated, can result in cancer.

➤ Genes that normally regulate the cell cycle or a cell's response to growth hormones—called *proto-oncogenes* in this context—mutate to form *oncogenes*.

➤ Other normal genes, called *tumor-suppressor genes*, exert negative control over cell division processes, and when they are mutated, they may fail to keep cell growth under control.

➤ The *accumulation* of mutated genes over an organism's lifetime can turn otherwise normal cells into cancerous cells.

iv. *Exposure to carcinogens and mutagens* are the most likely causes of the *somatic mutations* that contribute to cancer development, but mutagens may also cause germ cell mutations that are carried by gametes.

➤ *Carcinogens* are substances in the environment that increase the risk of cancer.

➤ Most carcinogens are mutagens.

➤ *Mutagens* are substances that cause mutations. Examples include tobacco, asbestos, x-rays, ultraviolet light, and a host of other chemical substances in the environment.

v. Some *viruses* can contribute to cancer development by transferring oncogenes to host cells, or causing mutations in proto-oncogenes or tumor-suppressor genes of host cells.

Test Tip

Know the names, descriptions, and effects of the different types of mutations.

Genetic Variation in Prokaryotes and Viruses

A. Prokaryotes reproduce *asexually*, but they also have several methods of genetic recombination.

1. *Transformation* occurs when a prokaryote takes up foreign DNA from its environment.
2. *Transduction* is when a virus transfers prokaryotic DNA from one cell to another.
3. *Conjugation* occurs when a plasmid is transferred from one prokaryote to another through a special tube-like structure called a *pilus*.

B. Because prokaryotes have a very short generation time, mutation and genetic recombination play important roles in producing and maintaining genetic diversity.

C. Virus replication involves invading a host cell and eventually living off of the host by taking over the metabolic machinery (parasitic).

1. Viruses cannot reproduce independently.
2. Viruses attach to the host via cell surface receptors and inject their DNA into the host.
3. During the *lytic cycle*, which is a type of viral reproduction, the virus eventually kills the host.
4. During the *lysogenic cycle,* the virus replicates its genome without killing the host and forms *prophage* (incorporation of the viral DNA into the host chromosome).

D. Viruses are efficient at rapid evolution and acquiring new phenotypes. The wide variety of possible hosts, and in turn, the enormous possibilities of chromosomes and their respective genes through which a virus can replicate itself all contribute to viruses' abilities to possess the potential for expansive genetic variations.

E. *HIV*, the causative agent of acquired immunodeficiency syndrome (AIDS), is a retrovirus; it uses an enzyme called reverse transcriptase to synthesize DNA from an RNA strand and infects T4 helper cells.

F. *Prion* is the protein infectious particle or misfolded protein that converts other normal proteins into mutant form; it is also the causative agent for "mad cow disease."

IV. Variation Due to Sexual Reproduction

A. Sexual reproduction produces genetic variation in three ways: independent assortment of homologous chromosomes, crossing over, or random fertilization.

B. The first two genetic recombination events occurs during meiosis I.

1. *Independent assortment of homologous chromosomes*—occurs during metaphase I and anaphase I, creating a variety of outcomes (gametes) that contain different combinations of an organism's maternal and paternal chromosomes.

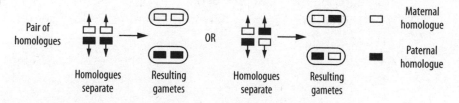

Depending on how homologues align during meiosis I, gametes with different combinations result.

2. *Crossing over*—occurs between homologous chromosomes during prophase I, creating entirely new chromosomes on which the organism's maternal DNA is mixed with his or her paternal DNA so that newly created chromosomes may be passed on to offspring.

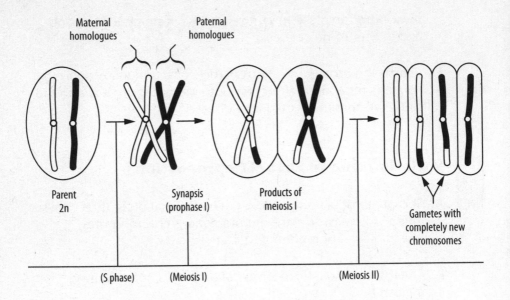

Maternal homologues   Paternal homologues

Parent 2n

Synapsis (prophase I)

Products of meiosis I

Gametes with completely new chromosomes

(S phase)   (Meiosis I)   (Meiosis II)

C. *Random fertilization* creates further genetic variation.

1. One sperm, out of the large variety of sperm a male can produce, joins with any of the large number of different eggs a female can produce.

2. This creates a large variety of different possible offspring.

Cell Communication

I. Key Concepts

A. Cell-to-cell communication is essential for multicellular organisms and their overall development, growth, and homeostasis.

B. Cell-to-cell communication between cells is also important for unicellular organisms.

C. Universal mechanisms of cell communication suggest an evolutionary similarity among species.

II. Evolutionary Similarities

A. Cells can communicate with each other in some of the following ways:

1. *Chemical messengers*—such as hormones
2. *Cell-to-cell contact*
3. *Synaptic signaling*—neurotransmitters diffuse across a synapse to a single cell

B. The three stages of cell signaling are *reception, transduction, and response.*

1. *Reception*—chemical signals bind to cellular protein.
2. *Transduction*—binding leads to a change along a signal transduction pathway.
3. *Response*—a specific cellular activity is triggered.

C. Signal transduction pathway—the process by which a signal on the cell's surface is converted to a specific cellular response—is strikingly similar in yeast and animal cells. Their evolutionary similarities in cell communication still exist today, despite the fact that the common ancestor of yeast and animals lived more than a billion years ago. Even signaling between bacteria and plants is similar in some ways.

III. Local- and Long-Distance Signals

A. Communication between cells involving *chemical messengers* can be *local* or *long distance*.

1. *Local regulators* are secreted by cells and only affect the activity of nearby cells.
   i. Examples include *neurotransmitters* in the nervous system, *histamine* in the immune system, *growth factors* in development, and *prostaglandins* in the immune and reproductive systems.
   ii. Local regulators act quickly and do not enter the bloodstream.
2. *Hormones* act over long distances because they enter the circulatory system and are transported around the body. Examples include insulin, testosterone, and estrogen, which are all regulated by the endocrine system.

B. When a hormone comes in contact with a *target cell*, it either enters the cell or binds to receptors on the surface of the cell.

1. Most chemical messengers, including most *protein-based hormones*, bind to *proteins* that act as *receptors* on the *plasma membrane* of target cells.
   i. The binding between a hormone and a receptor causes a physical change (usually involving movement) of the protein receptor, which sets in motion a message relay system inside the cell called a *signal transduction pathway*.
   ii. A series of *secondary messengers* carry the signal until it eventually results in a response by the cell.

iii. Common signal transduction pathways include *protein modifications* such as how methylation changes the signaling process, *protein phosphorylation*, activation of *G-proteins* and *cyclic AMP*, or increases in *calcium ion* levels.

iv. Changes in signal transduction pathways can alter cellular response, and in some cases where the pathway is blocked or defective, the changes can become deleterious, preventative or even prophylactic.

➤ An example would include an organism being bit by a poisonous spider that injects a type of toxin inside the organism's body; the poison blocks either a specific transduction pathway or a series of them. Effective treatment of the bite includes removal of the block to restore the health of the organism.

➤ Sometimes creating a block of a signal transduction pathway is the goal, so medication like anesthetics is used on the organism in order to block pain, for example.

2. The *fat-soluble steroids* and *thyroid hormones* pass through the plasma membrane and bind to receptor proteins within the cytoplasm or the nucleus of cells.

i. Hormone binding to a *cytoplasmic receptor* may trigger the response in the target cell.

ii. Some steroids *enter* the *nucleus* bound to DNA regulatory proteins and *stimulate transcription* of specific genes.

C. Cells can also *communicate cell to cell*. Examples include antigen-presenting helper T cells, Killer T cells, and plasmodesmata between plant cells that allow the transport of materials from cell to cell.

D. Signal transmission between cells can also affect gene expression. For example, *cytokines* can regulate gene expression by regulating cell replication and division, and *ethylene levels* signal changes in specific enzymes in fruit that indicate that it is time to ripen.

E. Communication between cells also affects cell function. For example, changes in the activity of gene *p53* can cause cancer.

Organismal Communication and Behavior

I. Organismal Communication and Behavior

A. Both *heredity* and *environment* influence behavior.

1. Like all phenotypes, the basis for behavior is genetic.
2. All behaviors require interaction with the environment because they are defined as a response to environmental stimuli.
3. Environment plays a large role in shaping the expression of most behaviors, and, as a result, the same *genotype* often results in a wide range of varying *phenotypes*.

B. *Innate behavior* is behavior that is *not learned*.

1. In response to a stimulus, called a *sign stimulus*, an organism may behave in a predictable fashion that does not need to be learned in order to occur.
   i. For example, males of many species demonstrate aggressive behavior or sexual behavior in response to the sign stimulus of red color.
   ii. Infants, when presented with even the simplest resemblance of a human face, smile.
   iii. Some hatchlings lift their heads and open their mouths toward the sky in response to any motion in their vicinity.
2. *Kinesis*—a type of innate behavior in which an organism lowers or raises its activity level depending on environmental conditions.
   i. For example, pillbugs move around less in dark areas as opposed to light areas.

        ii. The bug's movements are random, but the differences in activity levels make it more likely that the bug will find a dark location and stay there.

   3. *Taxis*—the innate movement toward or away from a stimulus.

        i. For example, when moths move toward light, it is called *phototaxis*; when mosquitoes move away from a repellant, it is called *chemotaxis*.

        ii. Movement toward a stimulus is a *positive taxis* (e.g., movement toward light is called positive phototaxis), and movement away from a stimulus is a *negative taxis* (e.g., movement away from light is called negative phototaxis).

        iii. Examples include the fight or flight response, protection of one's young, and avoidance responses.

C. *Learned behavior* involves modification of a behavior in response to experience.

   1. *Habituation*—a learned behavior involving the loss of responsiveness to a stimulus that occurs repeatedly without resulting in harm.

        i. An example is learning to be able to sleep with nonthreatening background noise, even though another set of noises results in waking.

        ii. Another example is when birds learn to ignore warning calls of members of their own species if those calls are repeated without being connected to a consequent threat.

   2. *Imprinting*—innate behavior that has a learning component that only occurs during a *critical period*—a specific time period, usually early in development, when the behavior is learned; the behavior is usually irreversible.

        i. For example, bonding between parents and offspring involves some type of recognition between the two that generally only occurs shortly after birth or hatching, such as when goslings bond with their mother in the first few hours of life and, thereafter, follow her around as opposed to any other adult bird.

        ii. If the behavior is not learned during the critical period, the behavior is usually *absent or aberrant* (different), such as when young birds do not hear their species' song during the critical developmental period, and thus fail to be able to learn to sing the song later in life.

3. *Associative learning* is learning to associate one stimulus with another.

  i. *Classical conditioning*—when an animal learns to associate an arbitrary stimulus with a reward or punishment.
    ➤ For example, a dog may learn to salivate in response to an unrelated stimulus, such as the ringing of a bell, if the ringing of a bell is always followed by a treat.
    ➤ Or a cat may learn to jump off the table when it hears a clapping sound, if the clapping sound has been associated previously with a punishment such as being sprayed with water.

  ii. *Operant conditioning* is closely related but involves an animal associating one of its own behaviors with a reward or punishment through *trial-and-error learning*.
    ➤ For example, a rat may learn to push one lever of a particular type that delivers a reward, while ignoring other levers that do not result in a reward by trial and error. Or a dog may learn to avoid all skunks after being sprayed by one.

D. *Cognition*—the ability of a nervous system to perceive, store, process, and respond to information obtained by the senses.

  1. A *cognitive map* is an internal representation or code of the spatial relationships among objects in an organism's environment.
    i. Migration may involve cognitive mapping.
    ii. Bees use cognitive mapping to remember and communicate the locations of food sources.

  2. *Consciousness* (awareness of self and environment) involves cognition.

E. *Social behavior*—the interaction between members of the same species (or sometimes members of different species), and includes aggression, courtship, cooperation, and deception.

  1. *Agonistic behavior* may occur when two members of the same species are competing for the same resource and

includes *ritual threatening* and ritual *submission* or, less often, actual fighting involving injury or death.

  i. Agonistic behavior can be the basis of *dominance hierarchies* where each member of a social group has a ranking within the group.

  ii. *Territories*—physical space partitions within a species' home range—are usually defended by agonistic behavior.

2. *Courtship*—another type of ritual behavior occurring before two members of the same species mate, and may involve agonistic behavior and/or assessment behavior.

  i. Agonistic behavior between males may be involved: a situation in which a competition determines which male will mate with female(s). Natural selection and survival of the fittest are seen in these courtship and mating behaviors.

  ii. *Assessment behavior* of females may be involved: a situation in which females choose between males displaying specific physical or behavioral characteristics highlighting their health or parenting abilities.

  iii. Mating systems can be *promiscuous* (no pair bonding), *monogamous* (one female and one male), *polygamous* (one male and multiple females), or *polyandrous* (one female and multiple males).

    ➤ The amount of *parental care* required may influence the type of mating system of a species.

    ➤ *Certainty of paternity* may influence the type of mating system as far as the investment of paternal parental care is concerned.

3. *Communication* is a feature of social behavior within a species that usually facilitates cooperation.

  i. Types of communication include *auditory* (vocalizations, or other sounds), *visual* (e.g., displays, dances, light flashes), and *olfactory* (e.g., *pheromones*).

  ii. The purpose of some communication may be *deception*, such as when certain fireflies flash the signals of other species and then eat the other species' members that respond to the signal.

4. *Altruistic behavior* may be the result of *inclusive fitness*; for example, an animal may behave in a way that decreases its

own fitness in order to increase the fitness of a *kinship group* that carries a large percentage of genes similar to its own.

i.   For example, a bird or mammal may give a warning call that directs the attention of a predator to itself, placing it in danger, but allowing related individuals to protect themselves.

ii. Altruistic behavior may be influenced by the *coefficient of relatedness*—the proportion of genes an individual shares with a particular member of its species—such that the more genes an animal shares with another member of the group, the more likely altruistic behavior is to occur.

➤ The coefficient of relatedness is 50% between a mother or father and his or her offspring, and also between full siblings.

➤ The coefficient of relatedness is 25% between an individual and his or her full blood-related aunts, uncles, nieces and nephews; and 12.5% between cousins.

➤ The coefficient of relatedness among all (female) worker bees within a hive is 50%.

iii. *Kin selection*—the evolutionary selection of behaviors that increase the likelihood of preserving a group of related individuals—is the hypothesized mechanism of inclusive fitness.

F. *Summary* of *core concepts* of animal *behavior.*

1. Heredity and environment contribute to most behaviors.

2. Behaviors can be innate or learned, or a combination of both.

3. Behavior between members of the same species involves communication of some sort.

4. Agonistic behaviors are used to settle disputes over resource partitioning between members of a species.

5. Courtship behaviors serve to bring two members of the same species together for sexual reproduction and may serve to maximize fitness of offspring.

6. Inclusive fitness may explain altruistic behavior among members of a kinship group.

PART V
INTERACTIONS

Population Dynamics

I. Key Concepts

A. Populations may experience exponential growth if there are no limiting factors in their environment and logistic growth if there are limiting factors.

B. The smaller a population is, the more likely it is to become extinct.

C. Worldwide, the human population is currently experiencing exponential growth, but is expected to begin to level off in the near future.

II. Individuals and Populations

A. Individual organisms have mechanisms to withstand physical changes in their immediate environments.

B. The environment of an organism includes biotic and abiotic factors.

1. *Biotic* factors include all living organisms within the environment.
2. *Abiotic* factors are the physical factors of the environment and include temperature, precipitation, humidity, wind, salinity, and availability of oxygen, nutrients, and sunlight.

C. As abiotic factors change over time, or from place to place in an organism's environment, an organism may respond in a variety of ways.

1. A *tolerance curve* describes how able or active an organism is over the range of change it may experience for a particular factor in its environment; at the extreme limits of its range, an organism may not survive.

Temperature Tolerance Curve

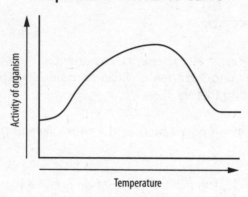

2. An organism may *acclimate* (adjust its tolerance) to an environmental factor, such as when humans produce more red blood cells as their bodies adjust to higher elevation.

3. *Regulators* are organisms that spend metabolic energy to internally regulate a physical factor, such as temperature or salinity, to keep it within a limited range even though their environment may exhibit a wider range for that factor.

4. In *conformers*, the factor is not internally regulated; instead, the conformer's internal factor changes to match the environmental factor as it increases or decreases.

Regulators and Conformers
Respond Differently to Environmental Change

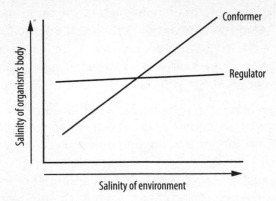

5. Individuals may respond to environmental change by temporary or permanent escape.
   i. An individual may move from location to location during the day, or *migrate* to another location.
   ii. An individual may become *dormant* through one of the following mechanisms.
      ➤ Some organisms may experience a period of inactivity called *torpor*: *hibernation* in cold weather or *estivation* in hot weather.
      ➤ Some organisms have resistant forms such as *spores* or *seeds*.

D. A *niche* refers to all of the roles of a species within its environment, including the biotic and abiotic features of its environment.

1. The *fundamental niche* of a species is the total range of environmental factors it can tolerate and the total range of resources it can potentially use.
2. The *realized niche* of a species is the actual extent to which it tolerates and uses its potential environment, due to the possibility that these resources are often reduced by biotic factors such as competition with other organisms.
3. *Generalists* are species with very broad niches, whereas the niches of *specialists* are more specific and limited.

III. Density and Dispersion

A. Population—group of individuals of the same species living together in the same location during the same period of time.

B. Population characteristics include size, density, patterns of dispersion, and age structure.

1. *Population size*—the number of individuals in the population; can be measured by direct counting in small populations or by sampling a portion of the population if it is larger.

2. *Population density*—refers to the number of individuals in a defined unit of space, such as the number of single-cell algae per milliliter of pond water or ferns per square kilometer of forest floor.

3. *Dispersion*—the pattern of distribution of individuals within a population.
   i. *Uniform* (or *even*) dispersion pattern—one in which the members of the population are spaced at relatively equal distances from one another, and often occurs in species that defend a defined territory.
   ii. *Random* dispersion pattern—each individual's position is independent of the locations of other individuals; for example, the dispersal of a plant's seeds by wind may result in the random location of the plant's offspring.
   iii. *Clumped* distribution pattern—the most common with most organisms in the population preferring to aggregate in the same area(s).
      ➤ Clumping can result from uneven distribution of the resources needed by the population's members.
      ➤ Clumping can also be the result of social behaviors that lead to swarming, flocking, or schooling among animals.

Dispersion Patterns

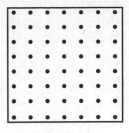
Uniform (even)

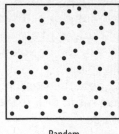
Random

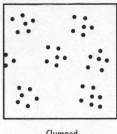
Clumped

4. The *age structure* of a population is dynamic and changes over time due to varying birth rates, death rates, and life expectancies.
   i. A population with a greater number of younger, reproductively active members is expected to increase in size more rapidly than a population with fewer young individuals.
   ii. If life expectancy increases, the number of older members in a population is expected to increase.

Age Structure in Two Different Populations

Population Increasing in Size

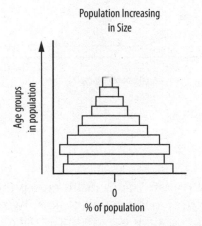

Age groups in population

0
% of population

Stable Population Size

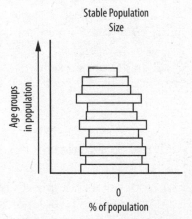

Age groups in population

0
% of population

5. A *survivorship curve* shows the expected mortality (death) rates of members of a population over their potential life span.

  i. A *Type 1* curve is characteristic of species that have few young and invest a lot of energy caring for them. Survivorship is high for early and midlife individuals, but drops precipitously with advanced age, indicating that most members of the population live out their potential, maximum life span.
  ii. *Type II* curves describe populations in which the members have more or less the same chance of dying regardless of age.
  iii. A *Type III* curve is characteristic of species that produce large numbers of offspring, most of which die before reaching maturity.

Age Structure in Two Different Populations

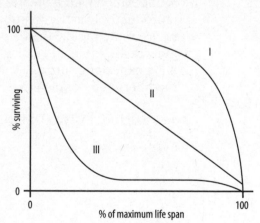

Type I—Death occurs for the elderly (i.e., humans)

Type II—Death is independent of age (i.e., fish or birds)

Type III—Death occurs for the young (i.e., plants or sea anemone)

IV. Growth Models

A. A population's growth rate (birth rate minus death rate) is the change in a population's size per a defined unit of time; two models are used to describe population growth under different conditions.

  1. *Exponential Model*—predicts the unlimited growth of a population because of no limitation on resources; the result is a J-shaped curve.

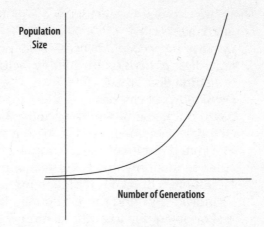

2. *Logistical Model*—more applicable to most populations in that it takes into account limited growth of a population due to limited resources; the result is an S-curve. The carrying capacity is the maximum population size that a habitat can hold (defined by the letter K).

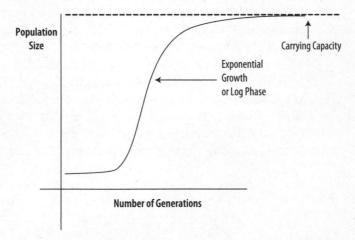

Be prepared to interpret population growth graphs and survivorship curves.

3. There are two types of factors that limit the growth of populations.
   i. *Density-independent factors*—affect population size regardless of density; most likely factors include weather or natural disasters.
   ii. *Density-dependent factors*—affect the population size based on the density of the population; most likely factors include food, predation, migration, or disease.
      ➤ Limited resources reduce population size in a density-dependent manner because competition for limited resources becomes more intense as the number of individuals using those resources increases.
      ➤ Poisoning due to accumulating waste materials becomes more likely, and affects more members of a population, as population density increases.
      ➤ Predation may be a density-dependent factor limiting a prey population if a predator increases its rate of predation when prey density is higher.
4. Small populations are more likely to become extinct than larger ones because inbreeding reduces the number, health, and genetic variability of offspring, or a local natural disaster could eliminate the entire population.
5. After remaining steady for most of human history, the human population has been increasing exponentially since the 1600s due to increasing life expectancy and greater ability to exploit resources, but the growth rate has slowed since the 1960s due to reduction in birth rate in developed, as well as many developing, countries.

Community Dynamics

I. Key Concepts

A. A community is a group of interacting populations of different species that live in the same geographic area.

B. Species richness, species diversity, and community stability are major characteristics of communities.

C. Species interactions and competition for resources are the bases for community relationships.

D. A succession of different communities occurs over time on newly created areas or in habitats destroyed by natural disasters or human activities.

II. Species Richness and Diversity

A. Species Richness—includes the number of different species in a community; Species Diversity—includes not only the number of each species, but the size of each population.

  1. Species richness increases as latitude decreases.
     i. Communities closest to the equator, such as those found in tropical rain forests, have the greatest number of species.
     ii. Three hypotheses may explain the greater number of species in lower latitudes.
        ➤ More available sunlight year-round promotes higher primary productivity—plant or phytoplankton

growth—resulting in a greater base level at the lowest trophic level of food chains.

➤ A more stable climate may lead to a greater number of niches available for exploitation.

➤ Tropical communities are older than those farther from the equator because they were not destroyed by recent ice ages.

Species Richness and Distance from Equator

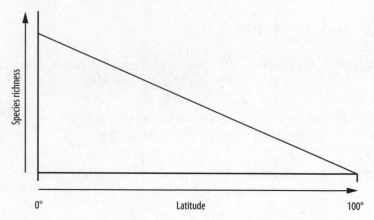

2. The *species-area effect* shows that species richness increases as the number of habitats and the area they cover increases. Therefore, larger islands have more species than smaller islands, and a reduction in habitat area is a main cause of extinction of populations.

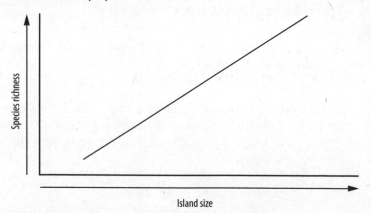

3. A *keystone predator* may increase species richness by preying on a successful competitor, thereby reducing competition between the prey species and its closest competitors and allowing those competitor populations to thrive.

Species Diversity with and without a Keystone Predator

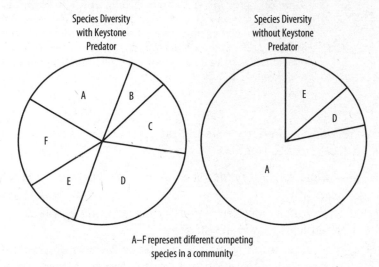

A–F represent different competing species in a community

B. *Communities with greater species richness are more stable* in the face of disturbances—such as droughts, floods, and other natural disasters—because a greater number of species tend to survive the disturbance in species-rich communities as opposed to species-poor communities.

III. Species Interactions

A. *Population Interactions*—interactions that occur with different species living in a community; can be beneficial to the species, but can also be detrimental to one or both of them.

B. *Symbioses*—close species relationships between a host and a symbiont

1. *Mutualism*—both species benefit from the relationship.

2. *Commensalism*—a relationship in which one species benefits while the other species is neither harmed nor helped.

3. *Parasitism*—beneficial to the parasite and detrimental—but not usually immediately fatal—to the host.

Test Tip

Symbioses may involve close coevolution where each species evolves in response to a change in the other species. Questions that require you to integrate material from different topics in biology are common on the test, so be aware of these relationships as you study; for example, ecology and evolution are closely related.

C. *Predation*—predator eats the prey; one species benefits while the other does not.

1. *Adaptations for predator*—claws, teeth, poisons, speed, eyesight.

2. *Adaptations for prey plants*—thorns in plants, plant chemicals that ward off prey.

3. *Adaptations for animals*—cryptic coloration or camouflage, aposematic coloration or bright colors that warn, or warning noises.

4. *Mimicry*—prey resembles another species.
    i. *Batesian mimicry*—a harmless species mimics a species that is dangerous to the predator.
    ii. *Müllerian mimicry*—two harmful species resemble each other and create a cumulative effect against a predator.

D. *Parasitism*—parasite lives off the host; one species benefits while the other does not (i.e., viruses, tapeworms, and mosquitoes).

E. *Competitive Exclusion Principle*—states that two species cannot survive in the same *ecological niche* (the sum of the total abiotic and biotic factors in an ecosystem) because they are competing for the same limited resources. Ultimately, neither species benefits from this interaction.

Succession

A. *Succession*—the gradual progression of different communities over time that occurs on virgin territory or in a habitat recovering from natural or manmade disturbances and involves species changing the environment over time.

B. *Primary Succession*—begins slowly and takes longer than secondary succession (hundreds to thousands of years) because it occurs in areas that have not supported life in the recent past.

   1. An area involved in primary succession may be newly exposed or newly formed rock, such as rock exposed by a glacier receding or islands produced by volcanic action.
   2. Autotrophic bacteria, algae, and lichens that grow on rocks are common pioneer species—the first species to colonize the area. Soil eventually will be produced to support plants and insects, and more gradually, larger animals and plants will migrate to the area and form a community.

C. Secondary Succession—occurs in areas where soil still remains and where communities used to exist but have been destroyed by disturbances such as fire, farming, and mining; this process can be completed within a year.

In succession, existing species change the environment, making it more favorable for other species that outcompete them over time as a result.

Ecosystem Dynamics

I. Key Concepts

A. An ecosystem includes groups of interacting communities and their environment; thus, it includes both *biotic* (living) and *abiotic* (nonliving) components.

B. A trophic level is an organism's nutritional position in a food chain.

C. Energy flows through an ecosystem in a one-way direction, from the sun to progressively higher trophic levels, passing along only about 10% of stored energy at each trophic level while traveling from lower to higher levels.

D. In contrast to a stream of energy running through an ecosystem, nutrients and water are recycled in an ecosystem.

E. Exponential growth of the human population has had effects on every aspect of the biosphere, from the global level to individual species.

II. Energy Flow

A. *Trophic Levels*—division of organisms in an ecosystem; energy flows through an ecosystem from lower trophic levels to higher trophic levels.

1. *Primary Producers*—comprise the first trophic level and are almost exclusively dependent on solar energy; they are

photosynthetic organisms, such as plants and blue green algae.

2. *Consumers*—organisms that comprise subsequent levels and that are ultimately dependent on producers for their energy needs
   i. *Primary Consumers*— herbivores or plant-eating organisms.
   ii. *Secondary Consumers*—carnivores that eat the primary consumers.
   iii. *Tertiary Consumers*—carnivores that eat other carnivores or organisms below them.
3. *Detritivores*—derive their energy from dead organisms or detritus (i.e., fungi and soil microbes); they are extremely helpful in recycling matter.

B. *Food Chains and Food Webs*—illustrate energy interactions between members of a specific community.

1. *Food chains*—are single energy pathways in which food is transferred from one trophic level to the next.
2. *Food webs*—an elaborate web of organisms feeding at more than one trophic level.

Example of a Food Chain (----▶) within a Food Web (——▶)

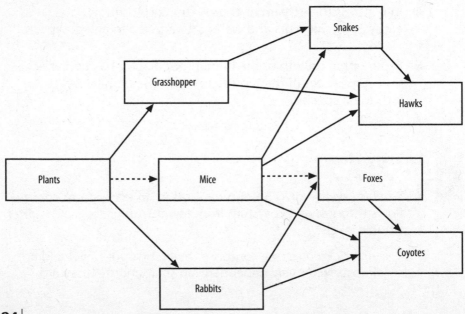

III. Nutrient Cycles

A. *Nutrient cycles* recycle water, carbon, nitrogen, and phosphorus, which move the ecosystem's organic matter to the abiotic portions of an ecosystem and back again.

B. *Water Cycle*—involves the processes of *evaporation, transpiration,* and *precipitation.*

C. *Carbon Cycle*—involves the processes of photosynthesis, cellular respiration, and combustion.

D. *Nitrogen Cycle*—involves a complex series of biochemical reactions by different soil bacteria to compounds assimilated by plants.

E. *Phosphorus Cycle*—provides the element phosphorus needed by all organisms as a component of nucleic acids and ATP.

Comparison of the Different Nutrient Cycles

Nutrient Cycled	Use in Organisms	Major Reserves	Important Processes
Water	Most of the mass of an organism is water	Bodies of water (oceans, lakes rivers, streams)	Evaporation Transpiration Precipitation
Carbon	Organic molecules	Organisms Atmosphere	Photosynthesis Respiration Combustion

(continued)

Comparison of the Different Nutrient Cycles (*continued*)

Nutrient Cycled	Use in Organisms	Major Reserves	Important Processes
Nitrogen	Proteins Nucleic acids	Organisms Atmosphere	Nitrogen fixation Ammonification Nitrification Assimilation Denitrification
Phosphorus	Nucleic acids ATP	Rock, soil Organisms	Weathering Decomposition

IV. Human Impact on Ecosystems

A. *Growth of the Human Population*—in recent times, the human population has increased significantly and has profoundly altered the biosphere from the global to the local level.

1. Human population has only risen to great numbers very recently (since approximately 1650 CE).

2. Humans have similar needs as other large animals, but their numbers, worldwide distribution, and unique ability to extract and utilize resources have a greater impact on the Earth than most other organisms.

   i. The use of fossil fuels to supply energy needed for industry and to heat and cool homes has resulted in a global increase in atmospheric carbon dioxide, which is correlated with a rise in global temperature and severe weather.

   ii. The disposal of biological, industrial, and household wastes have significantly altered nutrient cycling and introduced either totally new, or otherwise rare, toxic molecules into the environment that have significant adverse effects on humans and other species.

B. *Global Effects on the Ecosystem*

1. *Biological Magnification*—toxic chemicals being increased in concentration from one trophic level to the next. Biomass from one level is created from a larger biomass from the trophic level before. Top-level consumers are mostly affected. The best-known example is the use of DDT pesticide.

2. *Ozone Layer*—the ozone layer absorbs harmful UV light. CFCs or chlorofluorocarbons in aerosol cans and refrigeration units destroy the ozone by reducing it to oxygen.

3. *Greenhouse Effect*—carbon dioxide emissions from the burning of fossil fuels acts as a trap of solar heat in the atmosphere. Increases in carbon dioxide warm the air and accelerate the greenhouse effect. This is thought to be the major cause of global warming. Deforestation is also a major contributor to the greenhouse effect.

4. Despite these harmful effects on the ecosystem, some success has been achieved in reversing some of them.
   i. Thinning of the protective ozone layer of the atmosphere has been addressed.
   ii. An acid precipitation reduction plan has been implemented in some countries, including the United States.
   iii. Bans or reductions on toxins—such as DDT, PCBs, and mercury—have been achieved in some countries.
   iv. Climate change due to rising temperatures of the Earth has spurred development of models to predict the likely consequences and to provide information for making appropriate plans where possible.

Major Effects of Human Intervention on a Global Level

Problem	Cause	Global Consequence	Solution Attempted
Depletion of ozone layer surrounding the Earth	Use of CFCs	–More UV radiation –Greater cancer risk in humans –Greater likelihood of DNA damage in organisms	Phase out use of CFCs
Acid precipitation (pH below 5.6)	Sulfur and nitrogen oxides released from burning fossil fuels, especially coal	–Increase in aluminum toxicity in plants –Death of vulnerable organisms –Damage to entire ecosystems	Reduce sulfur and nitrogen oxides through "cap and trade" programs
Toxins, such as DDT, PCBs, and mercury	Industrial and household waste	–Poisoning and death of species in multiple ecosystems –Toxicity to humans	–Ban the use of the toxin –Clear industrial waste of toxin before release
Global climate change	Increased greenhouse gases, mainly from the burning of fossil fuels	–Polar ice melts, sea level rises –More and larger storms –Extinction of organisms sensitive to altered temperatures –Negative effects on agriculture	–Reduce greenhouse gases –Alternative energy sources –Plan for flooding, intense weather, and effects on agriculture

C. *Increased Rate of Species Extinction*—habitat destruction, overexploitation, and introduction of exotic species and diseases are the greatest threats to biodiversity and are currently contributing to an increase in the rate of species extinction.

1. *Biodiversity*—the degree of variation of species in a given area; the area can encompass a community or the entire biosphere.
2. Increasing *extinction rates* are currently decreasing biodiversity.
3. The major causes of decreasing biodiversity are over-exploitation, introduction of exotic species and diseases, and habitat destruction.
4. Reasons for preservation of biodiversity include utilitarian and nonutilitarian consideration.
5. *Conservation biology* focuses on maintaining biodiversity and includes strategies targeting all ecological levels.

D. *Sustainable Development*—limiting further damage to biodiversity and the environment involves sustainable development.

1. The goal of *sustainable development* is to manage the ecosystems of the biosphere in a way that supports the prosperity of human populations in the long term.
2. A continuing goal in this process is to study how ecological systems work in order to provide the best information for making decisions and how to manage and utilize the Earth's resources in a way that continues to replenish vital resources, such as clean air and drinking water, for future generations.

PART VI

THE EXAM AND
THE LABS

Science Practices and Essay Writing

I. General Tips on Essay Writing

A. *Overall Concept*—The AP Biology curriculum is unified by thematic underpinnings that can be related to any of the topics presented in this *AP Biology Crash Course.* A good way to utilize these themes is to incorporate them into your essays. The AP Biology readers will be delighted that you were able to see the big picture of the course rather than small isolated concepts. Below are the major themes:

Big Idea 1—The process of evolution drives the diversity and unity of life.

Big Idea 2—Biological systems utilize free energy and molecular building blocks to grow, to reproduce, and to maintain dynamic homeostasis.

Big Idea 3—Living systems store, retrieve, transmit, and respond to information essential to life processes.

Big Idea 4—Biological systems interact, and these systems and their interactions possess complex properties.

B. Tips for Essay Writing—Consider the following important factors when writing an essay for this exam:

1. Your Audience—
   It's crucial to remember "who" you are writing to. The graders of this exam are most likely individuals with knowledge of the subject matter and experience judging science essays. *They are interested in seeing what you know and how you think in a formal essay.*

2. Purpose of the Essay—

The purpose of the essay is to test your critical thinking skills, your writing skills, your ability to connect the "facts" and to reveal the big ideas.
   ➤ Always consider the bigger picture (i.e., the big ideas for this exam) and the implications of the facts and argument you are presenting to your readers.

3. Essay Prompt—

The first thing you should do when you start this portion of the exam is to read the essay question very carefully. It is crucial for you to understand exactly what it is asking you. A well-written essay that does *not* address the question will not help you. Underline or jot down specific key words and/ or statements in the essay prompt to help you remember the specific question(s) you must consider before writing the essay.

4. Prewriting—

Even though the clock is ticking and you may feel anxious about trying to produce a great essay in the allotted time, proper planning (i.e., a detailed outline) of your essay at the beginning of the session will benefit you greatly: (1) it helps you stay on track and not get off topic, (2) it allows you to see the "whole picture" of your argument before you start writing the essay, and (3) it is extremely useful as a guide to help you jog your memory as to what you're discussing during those moments of partial panic because time is almost up. So much time is wasted by many students when they look at the clock, begin panicking, and have a temporary brain freeze—causing them to forget what their last body paragraph was going to be about. If you have the outline in front of you, then just look at your paper and regain some confidence that you can produce a great essay in the allotted time.

5. Drafting the Essay—

At this point in the process, start writing your essay. You should have completed an outline of your main points and your thesis sentence during the prewriting phase. Follow your outline and answer the question in a formal essay format with an introduction, a thesis that answers the question, clear topic sentences that begin each body paragraph, analysis in each body paragraph with examples that are well explained, and a conclusion paragraph.

➤ Although it may be tempting to edit every word and sentence of your essay as you go, during this phase, just get the entire essay out. You can revise at the end. The first step is to finish the essay.

6. Revising the Essay—

This is the final step of the essay writing process and a crucial one. After you have written your essay, review the essay question again to refresh your memory as to the question your essay should answer. Then, first read your essay for content: Does it answer the essay prompt? Does your argument make sense? Then, reread your essay for grammatical and spelling errors.

II. Science Practices and Skills Required for the Exam

A. The following is a list of the science practices that the exam questions will test. You should use them as a guide to the overall skills that will be utilized by the exam questions; keep these science practices and skills in mind as you study the materials in this book. *Remember—the exam will heavily test your applied knowledge and will* not *test the plain memorization of facts.*

1. Use diagrams, graphs, supporting data, and models to communicate scientific phenomena and solve problems.
   Example: Analyze a model of DNA (see Chapter 14)
   Example: Analyze a tolerance curve graph (see Chapter 26)

2. Use mathematics appropriately. Some examples are:
   i.   Chi Square
   ii. Hardy-Weinberg Equilibrium
   iii. Mean and standard error deviation
   iv. Concentration gradient and osmotic potential
   v.   Rate determination
   vi. Solute concentration
   vii. Energy flow in ecosystems

3. Asking and answering scientific questions.

   Example: Articulate this question: "What is the evidence that supports natural selection?" and then be able to provide a specific and comprehensive answer.

4. Experimental Design and Data Collection
   i.   Independent variable
   ii. Dependent variable
   iii. Controlling other variables
   iv. Controls
   v.   Supplies and equipment
   vi. Experimental Protocol (method)

   The above points will be tested on the exam by asking you to justify the kind of data needed to arrive at a particular conclusion.

5. Data Analysis and Evaluation
   i.   Identify patterns and relationships
   ii. Explain what the data means
   iii. Evaluate whether data supports the hypothesis

   The above points will be tested on the exam by asking you to analyze specific data and the bigger picture. Also, you should be able to determine if the data provided is invalid.

6. Conclusions and Theories
   i.   Cite evidence to justify a claim
   ii. Use evidence to explain phenomena
   iii. Explain how theories are modified over time
   iv. Make predictions based on theories
   v.   Evaluate alternative explanations

   Example: Using the data provided to you to conclude that there is a population bottleneck and that this situation would change Hardy-Weinberg equilibrium.

   Example: Using specific genetic traits to determine phenotype.

7. Integration of Knowledge
   i.   Connect related information and concepts
   ii. Connect biology and other sciences

   Example: Analyze a phylogenetic tree in order to see the connections and ancestry and how natural selection and biodiversity have occurred over time.

The 13 AP Biology Labs

LAB 1   Artificial Selection

Analysis Question: How will you know if artificial selection has changed the genetic makeup of your population of plants?

Exercise 1A: Analyzing Plant Trichomes (1st Generation)

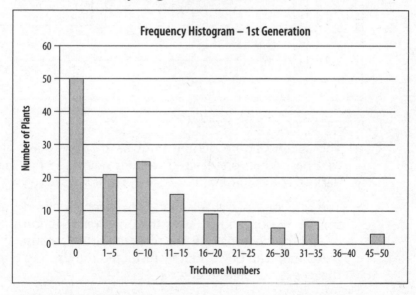

Interpretation of this Exercise:

➤ This is an exercise in understanding the characteristics of natural selection and artificial selection.

➤ This first histogram shows a selection of plants with the characteristic trichome, "plant hairs."

➤ You may be asked to determine if this figure is representative of natural selection or artificial selection. Specifically, this is an example of a random assortment of plants screened for a specific trait.

Exercise 1B: Analyzing Plant Trichomes (2nd Generation)

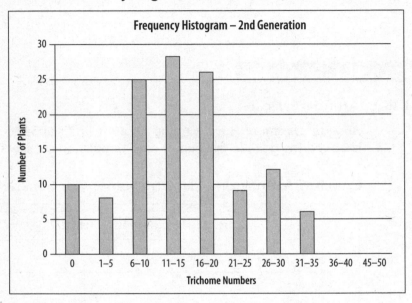

Interpretation of this Exercise:

➤ This second figure demonstrates a 2nd generation of plants that have been preselected to be grown based on having a desired characteristic (in this case, plant hairs).

➤ The goal of artificial selection is to favor a desired trait for a specific reason and is faster than natural selection at having organisms demonstrate this characteristic because the next generation can be absolutely restricted to offspring of parents that meet the desired criteria.

➤ In contrast, natural selection depends on the environment to do the selecting; therefore, traits like trichome would show up at a significantly slower rate than seen in two generations of plants.

➤ Questions on these figures may center on asking whether this figure demonstrates artificial selection or not, and why. How do you know for sure? The answer is: Yes, this is an example of artificial selection, and this is evident because the number of plants with the desired trait significantly increased only within one generation, which would not be the case if natural selection were the only selecting factor.

LAB 2   Mathematical Modeling: Hardy-Weinberg

Analysis Question: How can the use of mathematical models be used to investigate the relationship between allele frequencies in populations of organisms with evolutionary change?

Exercise 2A: Using Mathematical Modeling to Test for Hardy-Weinberg Equilibrium

p (the frequency of A) = 0.26						
q (the frequency of B) = 0.74					Tasters	Nontasters
		Gametes		Zygotes	AA	AB	BB
		B	A	BA	0	1	0
		B	A	BA	0	1	0
		B	A	BA	0	1	0
		A	A	AA	1	0	1
		A	B	BA	0	1	0
		A	A	AA	1	0	0
		B	B	BB	0	0	1
		B	B	BB	0	0	1
		A	B	AB	0	1	0
Genotype					2	5	3
					9 A's		11 B's

199

Interpretation of this Exercise

➤ The spreadsheet on the previous page represents one version of mathematical modeling that examines all the gametes, zygotes, and specific alleles within a random sampling of the population. The spreadsheet calculated the frequency of p (A) and q (B). The main ideas are that natural selection, as part of evolution, can act on a phenotype and create variations within a population. Evolutionary change is also driven by random processes, and populations of organisms continue to evolve.

Exercise 2B: Case Studies

CASE 1—A Test of Ideal Hardy-Weinberg

Interpretation of this Exercise

➤ A population that is in an Ideal Hardy-Weinberg would be a population of heterozygote individuals that follow all 5 key Hardy-Weinberg criteria:

- No mutation
- No gene flow or genetic variation
- A very large population sample
- No natural selection
- Random mating

    Frequency of p and q = 0.5

    Percent of p^2 = 25%

    Percent of 2pq = 50%

    Percent of q^2 = 25%

CASE 2—Comparison to Hardy-Weinberg

Interpretation of this Exercise

➤ If this population is random with no known restrictions on the population, can this population be in Hardy-Weinberg equilibrium?

- Frequency of p = 0.26 and q = 0.74
- Percent of p^2 = 10%
- Percent of 2pq = 35%
- Percent of q^2 = 55%

➤ Yes, it's possible for this population to be in Hardy-Weinberg equilibrium, even though the above calculations don't support this conclusion. The reason is that the population sampled is too small to reflect accurately whether there is Hardy-Weinberg equilibrium.

Exercise 2C: Uses of Mathematical Modeling for Hardy-Weinberg

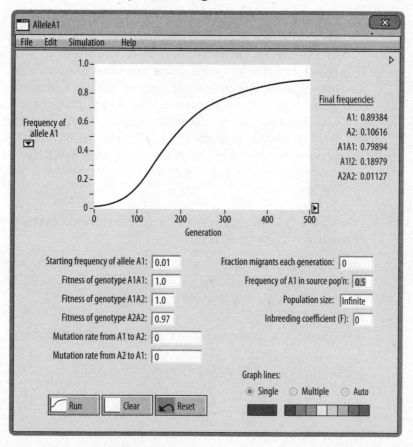

Interpretation of this Exercise

➤ The above figure shows an example of a program that can mathematically model the frequency of a specific allele by taking into account the starting frequency of that allele in a population and also the fitness of each of the genotypes. Once these numbers are applied to the program, then the program charts the increased frequency of this allele up to 500 generations.

➤ You may be asked to interpret the line chart on the previous page, which indicates that the frequency of allele A1 steadily increased its presence in the population over 500 generations, and at the 500th generation, its frequency in the population would be approximately 0.90.

➤ You also may be asked to explain the uses of this type of modeling. With the increased presence of computer programs able to articulate large amounts of data, programs such as the one above can quickly project the presence of specific alleles and take into account multiple factors that would alter its presence in one or more populations. These programs are useful learning tools for students to be able to plug in different allele frequencies and see the consequences on the chart; these programs are useful to researchers because they save time that otherwise would be spent calculating these projections by hand.

LAB 3   Comparing DNA Sequences to Understand Evolutionary Relationships with BLAST

Analysis Question: How can bioinformatics be used to help us better understand evolutionary connections and also the presence of genes in multiple organisms?

Exercise 3A: Constructing a Cladogram from Data in a Chart

Organism	Vascular Tissue	Seeds	Flowers
#1	0	0	0
#2	1	0	1
#3	1	1	1
#4	1	0	0
Total	3	1	2

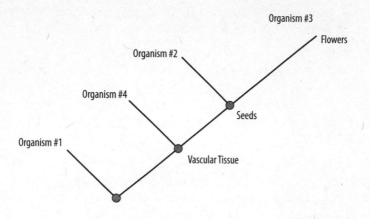

Interpretation of this Exercise

➤ Cladograms are used to map out the evolutionary connections between different organisms and also to demonstrate
which specific traits (e.g., vascular tissue, seeds, etc.) seem
to have evolved first in evolutionary history. This exercise is
meant to test your ability to articulate the data in the chart
and to draw the above cladogram. You also may be asked to
analyze a cladogram, like the one above.

Exercise 3B: Constructing a Cladogram from Gene Percent Similarities to Humans

Organism	Gene Similarity
#1	97%
#2	92%
#3	78%
#4	62%

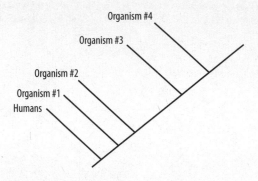

Organism #4
Organism #3
Organism #2
Organism #1
Humans

Interpretation of this Exercise

➤ The above chart shows the percentage similarity that each of the organisms #1–#4 are similar to humans. According to the above chart, organism #1 is the most similar, in reference to this gene, to humans. This exercise is meant to test your ability to articulate the data in the chart and to draw the above cladogram. This particular type of cladogram is meant to show the percentage similarities; therefore, the approximate placement of each of the branches is important.

➤ The program BLAST can also be used to determine how similar a specific gene, nucleotide, or protein is to other organisms, which includes humans, mice, etc. The database will be searched for the specified sequence and the results are listed by their percentage similarity.

➤ You may be asked to interpret data from a BLAST search that indicates the percentage similarities, just like the above chart. The skill is still the same: the higher the percentage similarity, the closer they should be mapped on the cladogram.

BIG IDEA 2: CELLULAR PROCESSES: ENERGY AND COMMUNICATION

LAB 4 Diffusion and Osmosis

Analysis Question: What causes plants to wilt if you forget to water them?

Exercise 4A: Watering Plants and Turgor Pressure

Interpretation of this Exercise

➤ Normally, when a plant has enough water supply, the cell has enough water to "feed" itself as well as enough water to store in its vacuoles; these vacuoles, full of water, are what help the leaves stay more stiff by creating turgor pressure. So when a plant does not get watered for quite a while and begins to wilt, the plant has begun drawing water out of the plant's system, but more important, out of its vacuoles containing water.

Exercise 4B: Diffusion

	Initial Contents	Solution Color		Presence of Glucose	
		Initial	Final	Initial	Final
Bag	15% glucose and 1% starch	Clear	Blue/ black	Yes	Yes
Beaker	H$_2$O & IKI	Yellowish	Yellowish	No	Yes

Interpretation of this Exercise

➤ Starch is too large a molecule to escape through the pores of the dialysis bag. As a result, the content of the dialysis bag turns from colorless to blue/black because the IKI from the beaker diffuses through the pores and reacts with the starch (a positive test for starch). The beaker fluid stays yellowish because no starch has diffused from the dialysis bag. Glucose is present in the beaker because it is a smaller molecule than starch and diffuses through the dialysis bag's pore. Residual glucose is still in the bag; thus, you continue to have a positive result for the bag.

Exercise 4C: Osmosis

Contents in Bag	Percentage Change
0.0 M Distilled Water	0.1%
0.2 M Sucrose	2.7%
0.4 M Sucrose	5.0%
0.6 M Sucrose	8.1%
0.8 M Sucrose	11.0%
1.0 M Sucrose	14.1%

% Change in Mass vs. Concentration of Sucrose

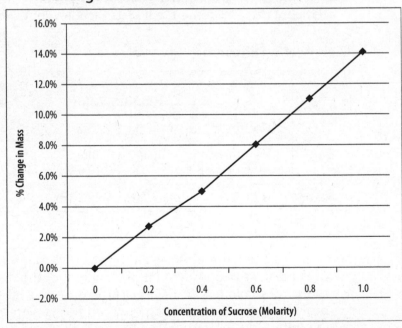

Interpretation of this Exercise

➤ As the concentration of the solute increases, water diffuses into the dialysis bag (hypotonic to hypertonic), increasing the mass of the dialysis bag. As an isotonic solution is evident with distilled water, diffusion is occurring at equal rates into and out of the dialysis bag.

Exercises 4D: Water Potential

Contents in Beaker	Percentage Change
0.0 M Distilled Water	19.0%
0.2 M Sucrose	8.0%
0.4 M Sucrose	– 5.0%
0.6 M Sucrose	–13.0%
0.8 M Sucrose	–21.0%
1.0 M Sucrose	–27.0%

% Change in Mass vs. Concentration of Sucrose

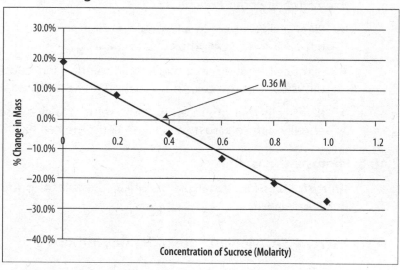

Interpretation of this Exercise

➤ The contents in the beaker and the percentage change in mass, as recorded in the chart above, is then graphed to demonstrate the relationship of the change in mass to the concentration of sucrose. Essentially, when the line crosses the x-axis at 0.36 M (estimation), then the concentration of the potato core is isotonic to the sucrose concentration. A net change of 0% at $x = 0.36$ M is the concentration that the water potential in the potato tissue is equal to the sucrose concentration (isotonic).

➤ Calculation of Water Potential

$$\psi_s = iCRT = -(1)\left(.36\frac{moles}{liter}H_2O\right)\left(0.0831\frac{literbar}{mole°K}\right)(295°K) = -8.8 \text{ bars}$$

If the calculated water potential is less than the water potential surrounding the bag, then water will flow into the bag (more solutes molecules inside the bag). If the calculated water potential is greater than the water potential surrounding the bag, water will flow out of the bag (less solute molecules inside the bag). Thus, water will flow from high to low water potential.

Exercise 4E: Plant Cell Plasmolysis

Interpretation of this Exercise

➤ Plant cells that are in a hypotonic solution will cause water to diffuse into the cell, thus creating a turgid cell.

➤ Plant cells that are in a hypertonic solution will cause water to diffuse out of the cell, thus creating a plasmolyzed cell.

➤ Plant cells that are in an isotonic solution will cause water to equally diffuse across the cell, making the cell flaccid.

LAB 5 Photosynthesis

Analysis Question: What factors affect the rate of photosynthesis in living leaves?

Exercise 5A: Median Rate of Photosynthesis

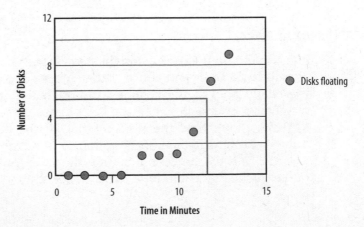

Interpretation of this Exercise

➤ Photosynthesis can be measured either by the production of O_2 or by the consumption of CO_2. The above chart shows how long it took each of the leaf disks to float to the surface—indicating that photosynthesis was taking place. The median of the disks floating is charted above, suggesting that it took approximately 12 minutes for half of the leaves to float to the surface.

Exercise 5B: Photosynthesis vs. Light Intensity

Light Intensity	Rate of Photosynthesis
0	0
200	0.05
400	0.06
600	0.07
800	0.08
1000	1.0
1200	1.1
1400	1.2

Interpretation of this Exercise

➤ The above chart shows that the increase of light intensity also increases the rate of photosynthesis. Be prepared to analyze this type of data either in chart form (as seen above) or on a line graph.

LAB 6 Cellular Respiration

Analysis Question: What factors affect the rate of cellular respiration in multicellular organisms?

Exercise 6A: How Temperature Affects Oxygen Consumption

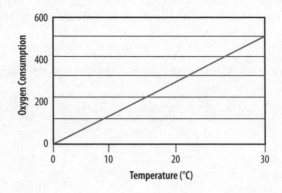

Interpretation of this Exercise

➤ The above line graph demonstrates that increased temperature is proportional to increased oxygen consumption. Be prepared to analyze data like that above and to compare the rates of cellular respiration for multiple organisms on the same graph.

BIG IDEA 3: GENETICS AND INFORMATION TRANSFER

LAB 7 Cell Division: Mitosis and Meiosis

Analysis Question: How do eukaryotic cells undergo mitosis or meiosis?

Exercise 7A: Observing Mitosis in Plant and Animal Cells Using Prepared Slides of Onion Root Tip and Whitefish Blastula

Interpretation of this Exercise

➤ Be able to draw a cell in Interphase (non-dividing portion of the cell cycle) and the 4 stages of mitosis.

Interphase

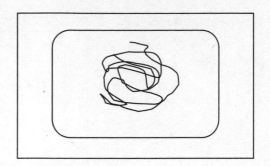

Prophase

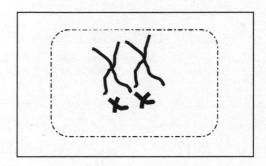

Metaphase

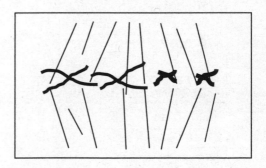

Anaphase

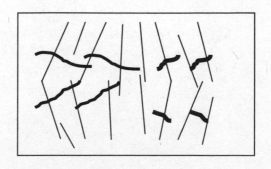

Telophase

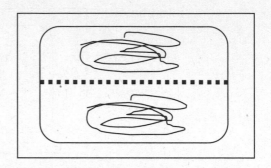

Exercise 7B: Time for Cell Replication

	Number of Cells				Percentage of Total Cells Counted	Time in Each Stage
	Field 1	Field 2	Field 3	Total		
Interphase	500	600	700	1800	82.0%	19 hr 41 min
Prophase	50	60	70	180	8.2%	1 hr 58 min
Metaphase	30	40	50	120	5.5%	1 hr 30 min
Anaphase	15	20	30	65	3.0%	43 min
Telophase	8	10	12	30	1.3%	19 min
				2195		

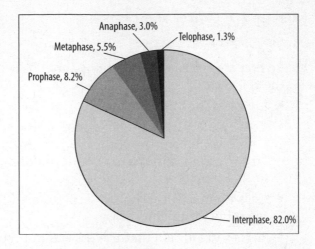

Interpretation of this Exercise

➤ The length of the cell cycle is roughly 24 hours with the majority of that time spent in Interphase, getting prepared for mitosis. Of the cells that one would observe in mitosis, the predominant phase is prophase.

Exercise 7C: Meiosis

Interpretation of this Experiment

➤ You must know the differences between meiosis and mitosis.

	Mitosis	Meiosis
Number of Chromosomes	2n-diploid	2n-diploid
Number of DNA Replications	1	1
Number of Divisions	1	2
Number of Daughter Cells Produced	2	4
Number of Chromosomes	2n-diploid genetically identical	n-haploid genetically variable
Purpose/Function	Growth of somatic cells	Generation of gametes

Exercise 7D: Cancer and Mitosis

Interpretation of this Exercise

➤ Be prepared to answer questions about how cancer can affect the cell cycle:

- Increases the rate of mitosis.
- Cells spent less time "checking" if everything is in order before continuing through the rest of the cell cycle; this "rush" causes mistakes to be made during replication and further mutations to daughter cells.
- Cells eventually can become so mutated that they do not resemble the original cells.
- Signals to indicate that the cell has not replicated properly and that the cell should undergo apoptosis (cell death) can be disregarded.

LAB 8 Biotechnology: Bacterial Transformation

Analysis Question: What are the ways we can utilize genetic engineering techniques to manipulate heritable information?

Exercise 8A: Bacterial Transformation

Plate Number	Condition	Observation
1	LB with transformed plasmid (positive control)	Lawn
2	LB without transformed plasmid (negative control)	Lawn
3	LB/Amp with transformed plasmid (experimental)	50 colonies
4	LB/Amp without transformed plasmid (positive control)	None

Interpretation of this Exercise

➤ Plate numbers 1 and 2 will have lawns (growth) of bacteria because there was no antibiotic in the plate agar.

➤ Plate number 3 had 50 transformed colonies because some of the cells were transformed with the plasmids containing the gene for resistance to ampicillin.

➤ Plate number 4 has no colonies since no plasmid was transformed, and the bacteria are susceptible to ampicillin.

Exercise 8B: Transformation Efficiency

$$\text{Total Efficiency} = \frac{\text{Total \# of colonies grown on agar plate}}{\text{Amount of DNA spread over agar plate}}$$

Interpretation of this Exercise

➤ Total mass of plasmid use = 0.0075 µg/µL x 20 µL = 1.5µg

➤ Total volume of cell suspension = 500 µL

➤ $\text{Mass of plasmid in suspension} = \frac{100\,\mu L}{500\,\mu L} = .2 \times 0.15\,\mu g = 0.03\,\mu g$

➤ $\text{Number of colonies per } \mu g \text{ of plasmid} = \frac{50 \text{ colonies}}{0.03\,\mu g} = 1.6 \times 10^2$

LAB 9 Biotechnology: Restriction Enzyme Analysis of DNA

Analysis Question: How can we use genetic information to profile individuals?

Exercise 9A: Restriction Enzyme Cleavage of DNA and Electrophoresis

Hind III

Actual bp	Measured Distance in cm
21,130	3.0
9,416	3.9
6,557	4.8
4,361	6.1
2,322	9.1
2,027	9.6
570	Cannot see on gel
125	Cannot see on gel

EcoR1

Band	Measured Distance in cm	Actual bp	Interpolated bp from Graph
1	2.8	21,226	19,000
2	4.4	7,421	9,000
3	4.9	5,804	7,000
4	5.1	5,643	6,800
5	5.7	4,878	5,000
6	6.9	3,530	4,300

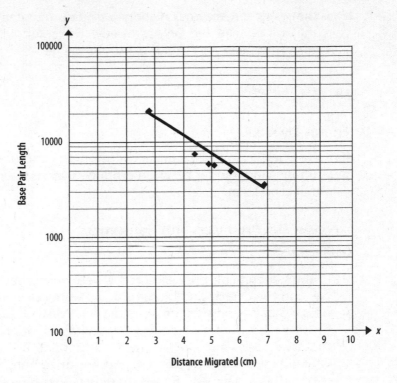

Interpretation of this Exercise

➤ Lambda phage DNA was incubated with restriction enzymes HindIII and EcoR1 separately. The migration distance of the DNA bands produced by HindIII were measured in centimeters and were plotted against bare paper size using semi-log paper. This was accomplished with DNA gel electrophoresis.

➤ Drawing the line of best fits allows for interpolation of the same DNA cut with EcoR1. Based on the line of best fit, the base pairs of the lambda DNA can be found and compared to the known value.

Important Points of this Laboratory

➤ Smaller pieces of DNA migrated faster and therefore are farther on the gel.

➤ The electrical current running through the buffer separates the DNA based on size. DNA is negatively charged; therefore, it migrates toward the positive end.

➤ If the restriction enzymes recognition site is mutated, the enzyme will not cut the DNA properly. The result will be the incorrect number and size of bands on the gel.

BIG IDEA 4: INTERACTIONS

LAB 10 Energy Dynamics

Analysis Question: What factors govern energy capture, allocation, storage, and transfer between producers and consumers in a terrestrial ecosystem?

Exercise 10A: Producers and Consumers

Interpretation of this Experiment

➤ In this experiment, brussels sprouts were fed to butterfly larvae and the energy and biomass flows were calculated at 12 days, 15 days, and after 3 days of growth. Most of the mass of the brussels sprout is water, which is an important product for the larvae to consume; therefore, it is important to understand why only fresh brussels sprouts, and not dried ones, must be used in this experiment. Be sure to understand an energy flow diagram and be able to draw one for this experiment.

Exercise 10B: Energy/Biomass Flow from Plant to Butterfly Larvae

Larva age (per 10 larvae)	12 days	15 days	3 days of growth
Wet mass of brussels sprouts	30 g	11g	19 g consumed
Plant percent biomass (dry/wet)	0.15	0.15	0.15
Plant energy (wet mass x percent biomass x 4.35 kcal)	19.58 kcal	7.5 kcal	10.56 kcal consumed
Plant energy consumed per larvae (plan: energy/10)	0.2 f	1.5 g	1.3 g gained

(continued)

Energy/Biomass Flow from Plant to Butterfly Larvae (*continued*)

Larva age (per 10 larvae)	12 days	15 days	3 days of growth
Wet mass of 10 larvae	0.15	0.15	0.15
Larvae percent biomass (dry-wet)	0.03 kcal	0.15 kcal	0.12 kcal
Energy production per individual (individual wet mass x percent biomass x 5.5 kcal/g)	0.03 kcal	0.15 kcal	0.12 kcal
Dry mass of the frass from 10 larvae	—	0.5 g	0.5 g excreted
Frass mass per individual	—	0.05 g	0.05 g excreted
Frass energy (waste) (frass mass x 4.76 kcal/g)	—	0.25 kcal	0.24 kcal excreted
Respiration estimate (plant energy consumed—frass waste energy production)	—	—	0.88 kcal

Interpretation of this Experiment

➤ According to the above chart, the wet mass of the brussels sprouts decreased over 15 days and eventually the larvae consumed 19 g of the brussels sprouts after growing for 3 days. The plant energy row demonstrates the transfer of energy from the plant to the larvae: the plant produces the energy for the butterfly to consume it. The plant energy consumed row reiterates this finding, and in fact, the wet mass of the larvae themselves increases as they consume the water content of the brussels sprouts. You want to familiarize yourself with this type of data chart and be able to read it and draw conclusions.

LAB 11 Transpiration

Analysis Question: What factors, if any, affect transpiration in plants?

Exercise 11A: Transpiration

Cumulative Water Loss in mL/m²

	Time (minutes)			
Treatment	0	10	20	30
Room	0	1.50	3.20	4.7
Light	0	4.00	8.12	12.13
Fan	0	4.21	8.45	12.30
Mist	0	1.50	2.00	2.33

Water Loss vs. Time

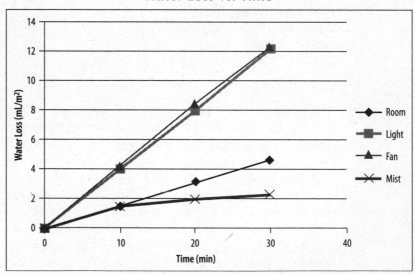

Interpretation of this Exercise

➤ Transpiration or the uptake of water from the leaf source is highest with both light and fan conditions. Both of these conditions cause water to be lost from the leaf surface. A water potential is created between the air surrounding the leaf and the photometer where the bottom of the steam contains water. Water will travel from an area of higher water potential to lower water potential. The mist condition mimics increased humidity, decreasing the water potential difference, since more water is occupying the surrounding air. The mist line is actually below that of the control (room) indicating the surrounding air has more water associated with it.

➤ The line graph above reconfirms the data in the chart: both the fan and light conditions increase transpiration over a 30 minute time period, and in fact, show relatively similar transpiration rates.

➤ The transpiration rate in the room is used as a reference point for the other conditions. For example, the mist condition shows that the transpiration rate is much slower than it would otherwise be in the room; this is also in stark contrast to the light and fan conditions.

➤ Another interesting conclusion that can be made is that the transpiration rate in the room continues to steadily rise over the 30 minute time period; however, the mist condition shows a slight increase of transpiration after 20 minutes had passed.

LAB 12 Fruit Fly Behavior

Analysis Question: What environmental factors affect fruit fly responses?

Exercise 12A: Environmental Factors

Environmental Factor	10 minutes	20 minutes
Salt	0	2
White Vinegar	25	60
Ripened Fruit	28	64
Sugar	10	18
Apple Cider Vinegar and Dish Soap	27	58

Interpretation of this Exercise

➤ The above chart shows how many fruit flies were present around or on the substance after 10 minutes and 20 minutes time. Since fruit flies are attracted to both overly sweet and vinegar-smelling substances, it is no surprise that the most flies were attracted to the white vinegar, the ripened fruit, and the apple cider vinegar with dish soap. The least amount of flies were attracted to the salt, and the flies probably were just checking out the substance briefly. Both the white vinegar and the ripened fruit were similarly attractive to the fruit flies. Although the sugar is obviously sweet, it was dry and crystalized and did not attract as many flies as the "sweet and wet" substances did. The apple cider vinegar with the dish soap also attracted about the same amount of fruit flies as the other vinegar and the ripened fruit; however, this substance also trapped most of the flies in the dish soap, while the apple cider vinegar attracted the flies to their deaths.

Exercise 12B: Reproductive Behavior in Fruit Flies

Interpretation of this Exercise

➤ Many organisms exhibit behaviors that indicate courtship. For *Drosophila melanogaster* a list of male and female characteristics are listed below:

• Male (tend to exhibit behaviors that promote mating): stamping the forefeet, circling the female, and wing vibration.

• Female (tend to exhibit behaviors that do not promote mating): ignoring, depressing wings, or flying.

Exercise 12C: The Life Cycle of *Drosophila*

Interpretation of this Exercise

➤ *Eggs*—small and oval shaped, and usually found on the side of culture tube.

➤ *Larval Stage*—wormlike stage that tunnel through the medium.

➤ *Pupal Stage*—fully mature larva are called pupa and tend to be brown in color. Basic body parts can be observed.

➤ *Adult Stage*—fly emerges from pupal casing and mating can take place again.

Exercise 12D: Crosses

Interpretation of this Exercise

➤ Cross 1 monohybrid

- Assume normal wings is dominant to dumpy (vestigial) wings. Cross a pure breeding long wing (W^+) to a dumpy (vestigial) wing (w).
- F_1 cross: W^+W^+ X ww → All progeny W^+w (all normal wings heterozygotes)
- F_2 cross: W^+w x W^+w → Progeny 1 W^+W^+: 2 W^+w: 1 ww (3 normal wings: 1 short wings)

➤ Cross 2 dihybrid

- Assume gray body color (g^+) and normal wings (w^+) is dominant to black body color (g) and dumpy (vestigial) wings (w). Cross a pure breeding gray and normal wing to black and dumpy (vestigial) wing.
- F_1 cross: $g^+g^+w^+w^+$ X ggww → All progeny $g^+g^+w^+$w (all gray, long-winged heterozygotes)
- F_2 cross: g^+gW^+w x g^+gW^+w → Progeny g^+gw$^+$w, ggww, g^+gww, ggw$^+$w
- (1:1:1:1 gray, normal wings; black, dumpy (vestigial) wings; gray, dumpy (vestigial) wings; black, normal wings)

➤ Cross 3 sex-linked

- Eye color is sex-linked in *Drosophila melanogaster*. Assume red eye (X^{w+}) is dominant to white eye (X^w). Cross pure breeding red eye female to a white eye male.
- F1 cross: $X^{w+}X^{w+}$ x X^wY → All progeny red eye: females are carrier $X^{w+}X^w$ and male X^{w+}Y
- F2 cross: X^{w+} X^w x X^{w+}Y → Progeny 1 $X^{w+}X^w$, 1 $X^{w+}X^{w+}$, 1 X^wY, 1 X^wY (all females have red eyes, ½ males have red eyes, and ½ males have white eyes).

Exercise 12E: Chi-Square Analysis

Interpretation of this Exercise

➤ Chi square is a statistical test to ensure the validity of a hypothesis.

- Null Hypothesis—there is no statistical difference between expected data and observed data.
- Alternative Hypothesis—another hypothesis that explains your observation.

➤ Formula is $X^2 = \sum \left[\dfrac{(o-e)^2}{e} \right]$

o = observed number of individuals

e = expected number of individuals

Σ = sum of values

Degrees of freedom = expected phenotypes -1

Use of the Chi Square Table of Critical Values

Probability (*p*)	Degrees of Freedom				
	1	**2**	**3**	**4**	**5**
0.05	3.84	5.99	7.82	9.49	11.1

If the calculated chi-square is greater than or equal to the critical value, the null hypothesis is rejected with a reassurance of 95%, meaning only 5% of the time would you see the null hypothesis as being correct.

Sample Data

Pheno-type	Number Observed	Number Expected	$(o-e)$	$(o-e)^2$	$(o-e)^2/e$
Normal wing	70	75	−5	25	0.33
Dumpy wing	30	25	5	25	1.00
				$X^2 = \sum \left[\dfrac{(o-e)^2}{e} \right]$	1.33

Result: There is no difference between observed and expected phenotypes; accept the null hypothesis.

LAB 13 Enzyme Activity

Analysis Question: How do abiotic and biotic factors influence the rates of enzymatic reactions?

Exercise 13A: Test of Peroxidase Activity

Interpretation of this Exercise

➤ The enzyme is peroxidase with the substrate being hydrogen peroxide (H_2O_2). The products released are water and oxygen gas.

$$Peroxidase + H_2O_2 \rightarrow 2H_2O_2 + O_2 (gas)$$

Note: Oxygen gas is flammable and will reignite a glowing flint.

➤ Note: Peroxide is a toxic byproduct of aerobic metabolism.

➤ Abiotic and biotic factors should affect the efficiency of this reaction.

Exercise 13B: Determining How pH Affects Enzymatic Activity

pH	3	5	6	7	8	10
	−0.002	0.543	0.321	0.160	0.056	0.004

Interpretation of this Experiment

➤ The above chart shows how pH affects the enzymatic activity of peroxidase. Its optimal pH environment, according to the chart, is around pH 5.0, with also decent activity continuing at pH 6.0. There is almost no activity at pH 7.0 and extremely little activity at pH 8.0 and pH 10.

Exercise 13C: Determining How Temperature Affects Enzymatic Activity

Temp	4°C	15°C	25°C	43°C	55°C	70°C	100°C
	0.102	0.163	0.234	0.308	0.274	0.156	0

Interpretation of this Experiment

➤ The above chart shows how temperature affects the enzymatic activity of peroxidase. Its optimal enzymatic activity is around 43 degrees Celsius, with good activity continuing at 25 and 55 degrees Celsius. The far ends of the activity spectrum that still include enzymatic activity are 15 and 70 degrees Celsius, with an even lessened activity at 4 degrees Celsius. The only absence of enzymatic activity occurred at 100 degrees Celsius because the enzyme became denatured and is unable to function properly.

Notes

Notes

Notes

Notes

Notes

Notes

Carolyn David if found, call 203-326-1214